Charles Wesley Purdy

Bright's Disease

And Allied Affections of the Kidneys

Charles Wesley Purdy

Bright's Disease
And Allied Affections of the Kidneys

ISBN/EAN: 9783337155919

Printed in Europe, USA, Canada, Australia, Japan

Cover: Foto ©berggeist007 / pixelio.de

More available books at **www.hansebooks.com**

BRIGHT'S DISEASE,

AND

ALLIED AFFECTIONS OF THE KIDNEYS.

BY

CHARLES W. PURDY, M.D. (Queen's Univ.),

HONORARY FELLOW OF THE ROYAL COLLEGE OF PHYSICIANS AND SURGEONS, KINGSTON; MEMBER OF THE
COLLEGE OF PHYSICIANS AND SURGEONS OF ONTARIO; PROFESSOR OF GENITO-URINARY
AND RENAL DISEASES IN THE CHICAGO POLYCLINIC; MEMBER OF THE
CHICAGO ACADEMY OF SCIENCES; MEMBER OF THE ILLINOIS
STATE MICROSCOPICAL SOCIETY, ETC. ETC.

WITH NEW AND ORIGINAL ILLUSTRATIONS.

PHILADELPHIA:
LEA BROTHERS & CO.
1886.

TO

HENRY W. DAY, M.D.,

EX-PRESIDENT OF

THE COUNCIL OF THE COLLEGE OF PHYSICIANS AND SURGEONS OF ONTARIO,

THE FOLLOWING PAGES ARE AFFECTIONATELY

Inscribed

BY HIS FORMER PUPIL,

THE AUTHOR.

PREFACE.

THE object of this work is to furnish a systematic, practical, and concise description of the pathology and treatment of the chief organic diseases of the kidneys associated with albuminuria, which shall represent the most recent advances in our knowledge on these subjects.

The lack hitherto of systematic descriptions, both of scarlatinal and puerperal nephritis, has induced me to give these subjects the prominence in this work to which their great practical importance so justly entitles them; and while the chapters devoted to the consideration of these two subjects are by no means claimed to be exhaustive, yet it is hoped that they may constitute something more systematic and definite than has thus far been within reach of the busy practitioner.

In the study of renal pathology the anatomical divisions of nephritis, heretofore so generally employed, seem to me to be misleading; and, moreover, since they are, strictly speaking, incorrect, their use has been avoided throughout this volume.

My acknowledgments are due to Professor D. J. Hamilton, who not only afforded me the liberty of the Pathological Laboratory in the University of Aberdeen for the purpose of special investigation, but also so generously contributed much of his valuable time to the same end.

*

PREFACE.

My thanks are also due Dr. Frank Cary, of Chicago, for his faithful execution of the drawings from morbid anatomy with which this volume is illustrated.

Lastly, I am indebted to Dr. Joseph Matteson, of Chicago, and to Dr. Wm. H. German, of Morgan Park, for valuable assistance in revising the proofs, as well as for the interest they have manifested in the preparation of this work.

THE AUTHOR.

163 STATE STREET, May, 1886.

LIST OF ILLUSTRATIONS.

FIG.		PAGE
1.	Pocket-case	46
2.	Acute nephritis. Transverse section through cortex .	90
3.	Acute nephritis. Section of cortex	91
4.	Chronic nephritis. Longitudinal section of medulla near papilla	. 113
5.	Chronic nephritis. Section of cortex	. 115
6.	Sphygmographic tracing in chronic nephritis	. 138
7.	Cirrhosis of the kidney. Longitudinal section of cortex, including capsule	. 148
8.	Diagram showing method of gauging high arterial pressure .	. 154
9, 10.	Sphygmographic tracings of normal pulse	. 155
11–13.	Sphygmographic tracings from cases of cirrhosis of kidney	. 155
14, 15.	Sphygmographic tracings from a case of cirrhosis of kidney	. 176
16.	Scarlatinal nephritis, fourth day	. 197
17.	Scarlatinal nephritis, third week	. 199
18.	Lardaceous degeneration of kidney ; cortical section	. 254

CONTENTS.

CHAPTER I.

ALBUMINURIA.

PAGE

History—False Albuminuria—True Albuminuria and Causes—Significance of Albuminuria—Renal Casts—TESTING FOR ALBUMIN IN THE URINE: Urates; Phosphates; Mucin; Peptones; Foreign Substances in Urine; Heat; Nitric Acid; Picric Acid; Potassio-mercuric Iodide; Ferrocyanide of Potassium—HYGIENIC TREATMENT OF ALBUMINURIA: Diet; The Skin; Exercise; Psychical Influences; Climate 17–57

CHAPTER II.

URÆMIA.

ETIOLOGY—SYMPTOMS: Chronic Uræmia; Acute Uræmia—DIAGNOSIS —PROGNOSIS—TREATMENT 58–85

CHAPTER III.

ACUTE NEPHRITIS.

ETIOLOGY: Exposure to Cold; Febrile Disease; Toxic Agents; Cutaneous Lesions—MORBID ANATOMY: Microscopic—SYMPTOMS: The Urine; Dropsy; Digestive System; The Circulatory System; The Nervous System; The Respiratory System; Course and Duration—DIAGNOSIS: Constructive; Differential—PROGNOSIS—TREATMENT 86–108

x   CONTENTS.

CHAPTER IV.

CHRONIC NEPHRITIS.

PAGE

ETIOLOGY: Scarlatina; Pregnancy; Cold; Chronic Suppuration; Malaria; Syphilis; Alcohol—MORBID ANATOMY—SYMPTOMS: The Urine; Dropsy; The Circulatory System; The Muscular System; The Digestive System; The Nervous System; Course and Duration—DIAGNOSIS: Constructive; Differential—PROGNOSIS—TREATMENT: The Early Stages; Late Stages . . 109-139

CHAPTER V.

CIRRHOSIS OF THE KIDNEY.

ETIOLOGY: Age; Habits; Psychic Influence; Gout; Dyspepsia; Heredity; Plumbism; Climate, Malaria, etc.—MORBID ANATOMY: Microscopic—SYMPTOMS: Early Symptoms; Dyspepsia; Neuralgia; Vertigo; The Urine; Circulatory System; The Heart; Epistaxis; Mature Symptoms: The Digestive System; The Urinary System; The Respiratory System; The Vascular System; The Muscular System; The Cutaneous System; The Nervous System; Uræmic Amaurosis; Albuminuric Retinitis; Late Symptoms; Course and Duration—DIAGNOSIS: Constructive; Differential—PROGNOSIS—TREATMENT: Early Stage; Stage of Cardiac Enlargement; Advanced Stage . . . 140-191

CHAPTER VI.

SCARLATINAL NEPHRITIS.

ETIOLOGY: Predisposing Causes; Exciting Causes—MORBID ANATOMY—SYMPTOMS: The Urine; Dropsy; The Digestive System; The Circulatory System; The Respiratory System; The Nervous System; Course and Duration—DIAGNOSIS: Constructive; Differential—PROGNOSIS—TREATMENT: Prophylactic; Curative   192-225

CHAPTER VII.

PUERPERAL NEPHRITIS.

ETIOLOGY—MORBID ANATOMY—SYMPTOMS: Dropsy; The Urine; The Digestive System; The Circulatory System; The Nervous System; Course and Duration—DIAGNOSIS: Constructive; Differential—PROGNOSIS—TREATMENT 226-249

CONTENTS. xi

CHAPTER VIII.

LARDACEOUS DEGENERATION OF THE KIDNEYS.

PAGE

ETIOLOGY: Chronic Suppuration; Tuberculosis; Syphilis—MORBID ANATOMY—SYMPTOMS: The Urine; Dropsy; The Digestive System; The Circulatory System; The Glandular System; The Nervous System; The Respiratory System; Course and Duration — DIAGNOSIS: Constructive; Differential — PROGNOSIS — TREATMENT 250–270

CHAPTER IX.

CYANOTIC INDURATION OF THE KIDNEY.

ETIOLOGY—MORBID ANATOMY—SYMPTOMS: The Circulatory System; The Respiratory System; Dropsy; The Urine; Course and Duration—DIAGNOSIS: Constructive; Differential—PROGNOSIS— TREATMENT 271–280

BIBLIOGRAPHY . 281

BRIGHT'S DISEASE

AND

ALLIED AFFECTIONS OF THE KIDNEY.

CHAPTER I.

ALBUMINURIA.

ALBUMIN was first discovered in the urine by Cotugno, and the results of his observations were published in Vienna in 1770. He found that the urine of certain dropsical persons coagulated upon being boiled, but he did not suspect an abnormal condition of the kidneys as the cause of the albuminuria. Wells was probably the first to point out that certain anatomical changes in the kidneys, especially thickening of the cortex, were accompanied by albuminuria during life; but even he did not consider renal change essential to the production of albuminuria. Our own countryman, Blackhall, in 1820, called attention to the frequent coincidence of pathological kidneys with dropsy, and the appearance of albumin in the urine. But Blackhall looked upon both the albuminuria and dropsy, not as a result of diseased kidneys, but rather as a result of changes in the organism at large.

Richard Bright, of London, was the first to point out, in

1827,[1] that certain alterations in the kidneys first described by himself, were associated with albumin in the urine as a direct result, and he further claimed that albuminuria might be accepted as the index of these pathological conditions. Bright's observations were soon confirmed by others, and especially by Christison; although it was not long before objections were raised against his doctrine. Prout, especially, claimed that albuminuria was not caused by structural changes in the kidneys, but rather arose from abnormal states of the blood which permitted its serum to pass off by the kidneys.

It would not be in keeping with the objects of a practical work, such as this is intended to be, to enter into a minute history of the various theories which have from time to time been advocated as to the cause of albuminuria, much less to enter into a discussion of them. Such conclusions only as seem at present fairly well established will be put forward in these pages.

In considering the subject of albuminuria from a general standpoint, it may, perhaps, be most practically dealt with under two divisions: (*a*) extra-renal, or false, albuminuria; and (*b*) renal, or true, albuminuria.

FALSE ALBUMINURIA.—Extra-renal, or false, albuminuria comprises those conditions in which albumin gains access to, and becomes mingled with, the urine after the latter leaves the kidneys—the urine being perfectly normal at the time it is secreted. In false albuminuria the proteid found in the urine is not serum albumin, derived directly from the blood, as in true albuminuria; but it is rather, for the most part, the product of pus, or cellular elements, which result from inflammatory or ulcerative changes

---

[1] Reports of Medical Cases selected with a view of Illustrating the Symptoms and Cause of Diseases by a reference to Morbid Anatomy, London, 1827.

taking place in some part of the urinary tract external to the kidneys.

As pus plays so important a part in extra-renal albuminuria, it may be well to review its chief characteristics, together with the various methods of production of albuminuria due to suppurative changes in the genito-urinary tract, apart from the kidneys themselves.

Pus corpuscles are made up of cellular elements and a fluid known as pus serum, which latter is mostly albuminous in character, and does not differ essentially from the albumin of the blood serum. Now cellular elements are insoluble in water, or in saline solutions like the urine, and hence when pus becomes mingled with urine it renders the latter turbid. *We, therefore, find the urine more or less turbid in extra-renal albuminuria.* The amount of albumin is comparatively small in cases in which it arises from disease of the genital tract, unless the inflammation giving rise to the pus be severe. In the latter case, the pus, being of greater gravity than the urine, settles to the bottom of the vessel as a greenish-yellow mass; and if the supernatant urine be poured off, and liquor potassæ added to the deposit, the latter forms with it a gelatinous, sticky mass, quite characteristic (Donne's test).

Pus corpuscles are best studied under the microscope, where their appearance is found to vary with the concentration of the urine, its reaction, and the quantity of salines contained therein.

Vogel pointed out that pus corpuscles are seen in two forms in acid urine—(*a*) globular, and (*b*) irregular, sending out processes.

In urine of low specific gravity (watery urine), or in alkaline urine, the corpuscles swell up two or three times larger than their normal size, and their granulation clears away, bringing out the nuclei more distinctly. If the suppuration producing the pus be of a healthy character, the pus corpuscles are more likely to preserve their regular outlines ; and, on the other hand, if the corpuscles appear irregular, sending out points or processes, it indicates the existence of some chronic ulcerative process of an unhealthy character.

A comparison of the number of pus corpuscles observed under the microscope, with the quantity of albumin contained in the urine, will usually lead to a correct conclusion as to whether or not the albuminuria be due alone to the presence of pus.

Circumstances will frequently throw light upon the nature of the albuminuria, and lead to the discovery of its true cause. If, for instance, albuminuria comes from pus formed in the urethra, or external to the bladder, the first ounce of urine passed will be turbid and albuminous, while that following will not only be clear, but also free from albumin.

In cases, too, of urethral inflammation, evidences thereof will usually be apparent at the meatus, if pus is not actually present there or on the linen. It is well to bear in mind that, in cases of female patients suffering from leucorrhœa, sufficient pus may gain access to·the urine to give it a slight albuminous reaction. In such cases a vaginal douche should be used previous to passing a sample of urine for examination.

Purulent catarrh of the bladder furnishes elements, which, on mingling with the urine, give the latter an albuminous reaction, which may be taken for true albuminuria. It is to be distinguished as follows : in vesical catarrh the urine usually has an ammoniacal odor ; it is alkaline in reaction, either when passed, or very soon after, as it quickly undergoes decomposition, thereby liberating free ammonia. The sediment is sticky and gelatinous, and contains pus corpuscles, bacteria, crystals of ammonio-magnesian phosphate, and vesical epithelium. There is usually present more or less pain above the pubis, with increased frequency of micturition, and in some cases the latter symptom is very urgent.

Pyelitis is distinguished from renal albuminuria as follows: in the former there is more or less pain and tenderness in the loins; pus, also, is present in the urine in greater or less quantity; and there may be tumor, and intermittent discharge of pus. There is usually some fever with pyelitis; sometimes rigors; and a previous history of renal colic is frequent. The absence of renal casts, especially if the urine be acid, will exclude the kidneys as the source of the albuminuria. In chronic inflammatory conditions of the prostate gland, more or less discharge is furnished of an albuminous character, which may give the urine an albuminous reaction. The presence of prostatic cylinders, amyloid bodies, and Bottcher's crystals,[1] points to the prostate as the source from which the albumin is derived.

It has been claimed that spermatic fluid, or spermatozoa, when mixed with urine, give the latter an albuminous reaction. According to my own observations, picric acid or potassio-mercuric iodide solution causes a copious precipitate in urine containing spermatic fluid; but this precipitate is redissolved by heat. No change occurs with heat alone, nitric acid, or ferrocyanide of potassium solution with citric acid: therefore, the spermatic fluid contains no serum albumin.

Finally, various, more or less circumscribed, morbid processes in the urinary tract external to the kidneys, such as ulcerations, inflammatory patches, small glan-

---

[1] Spermatic crystals are transparent and rhomboidal in form, composed of phosphoric acid, with an albuminous base (Belfield).

dular abscesses, cancer, tubercle, neoplasms, etc., may furnish elements, such as blood, pus, or cellular débris, or all combined, which may give the urine an albuminous reaction; and, unless carefully considered, may be taken for true or renal albuminuria. These various products are best studied by means of the microscope, which, together with the suggestions already made, and attending symptoms, will usually lead to a correct conclusion in the matter.

TRUE ALBUMINURIA.—Renal, or true, albuminuria deals only with albumin directly excreted by the kidneys with the normal constituents of the urine. The earlier writers were aware of but one form of albumin that thus appeared in the urine, viz., serum albumin of the blood; indeed, some writers, notably Prout, used the term serous urine instead of albuminuria. We now know that two forms of albumin derived from the blood are found in the urine under some circumstances, viz., serum albumin and globulin. Up to the present time the clinical significance of serum albumin only has received careful study; although so frequently associated with globulin. Globulin being the more diffusible of the two native albumins, we would expect it to escape from the blood the more readily in diseases of the kidneys. This is found to be true in some cases at least—globulinuria preceding serinuria for a longer or shorter time.

Furthermore, Senator has shown very recently that in nearly all cases of albuminuria, both serum albumin and globulin are present in the urine. As yet, however, sufficient attention has not been given to the subject of globulinuria to determine in all cases its clinical significance, but with more discriminating tests which are now being employed, it is not unlikely to assume considerable importance in the near future.

Serum albumin and globulin, then, are the only native albumins of the blood so far as we at present are aware; and they are both liable to pass into the urine in certain diseased conditions of the renal organs.

In addition to these, it is now known that other proteids may find their way into the urine from various sources. These are, for the most part, albumins in a transition state, from the untransformed proteid in the stomach in various stages, up to its perfect elaboration into serum albumin of the blood on the one hand; on the other hand, they represent a state of retrograde metamorphosis between serum albumin and suppurative necrosis, and are found in conditions approaching the latter.

Peptone is the most notable member of this group. None of these proteids exist in the blood in normal conditions of that fluid—at least not in appreciable quantities; and if through any means they gain access to the blood, they are quickly eliminated through the kidneys, and appear in the urine. Peptone, for instance, may from incomplete transformation in the process of digestion pass into the blood as such, and if so, it is excreted by the kidneys, resulting in peptonuria. This may result from ingestion of more albuminous food than the digestive organs can transform within a given time, and hence there is an overflow. Failure or impairment of duodenal digestion is believed to be a cause of this sometimes. Dr. Oliver believes (*Bedside Urine Testing*, 3d edition, 1885) that this failure of peptone transformation may arise through failure of the bile derivatives to fix (precipitate), as they do normally, the peptones; as a consequence, the latter are taken up unchanged into the blood, and separated through their more ready diffusibility by the renal glomeruli.

Albumin of the blood or tissues may, in consequence of certain morbid changes resulting from inflammation, infective fevers, etc., become subject to alterations through which as one of the results peptone is formed as a retrograde step of metamorphosis, and if so, it is thrown out by the kidneys, and peptonuria is the result. That such alteration is conceivable, as Senator writes (*Albuminuria in Health and Disease*), "appears more clear from the fact that we are taught by recent observations to the effect that no such sharp distinction exists between albuminous substances proper, and peptones, as we were formerly disposed to believe . . . .; recent investigations have likewise shown that the conversion of albuminous substances into peptone takes place gradually, and in stages, certain intermediate or transitional products being formed, and at least one such product is clearly recognized, analogous in some respects to albumin, and in others to peptone, viz., hemi-albumose."

From the fact that these intermediate modifications of albumin possess a clinical significance which points rather to morbid conditions outside of, than in the kidneys themselves, they do not demand more minute consideration here.

CAUSES.—It is generally accepted that renal albuminuria arises from four causes: (*a*) alterations in the blood-pressure, (*b*) changes in the vascular tunics, (*c*) degeneration of the renal epithelium, (*d*) alterations in the composition of the blood. The slighter degrees of albuminuria are often due to alterations in blood-pressure, such as increase of the arterial flow. As such conditions depend chiefly upon circumstances which are often transient, albuminuria from this cause is often transient as well as slight. Muscular exercise calls for increased cardiac power, which in turn raises arterial tension even twenty to twenty-five per cent. if the exercise be active; and it is not unusual to observe a slight amount of albumin in the urine of perfectly healthy persons undergoing active muscular exercise. In some persons, in consequence of abnormally forcible cardiac contractions, or other conditions which may disturb the balance in the renal glomeruli upon which normal filtration of urine depends, even passive movements may bring about slight or even considerable albuminuria. Under such circumstances albumin may be found in the urine during the day when the individual is about, while at night the urine may be free from albumin.

I believe it is under such circumstances that "cyclic albuminuria"—so called by Dr. Pavy—arises. (See *Brit. Med. Journ.*, Oct. 14, 1885.) This seems even more probable since Dr. Pavy himself has observed in some of these cases a sharp and unduly forcible cardiac impulse.

Heat, also, causes a direct rise in the arterial blood-pressure, as shown by increased tension of the pulse and turgescence of the skin. It is altogether likely that the transient albuminuria which occurs at the height of simple pyrexia is largely due to this cause.

Senator (op. cit.) has rendered this clear by the following experiments. He introduced rabbits into an oven constructed for the purpose, having previously drawn off their urine and tested it for albumin. He then slowly increased the heat in the chamber and again tested the urine: "the result in all cases was the production of albuminuria when the bodily temperature had been increased by 1.5° to 3°. C. with sufficient rapidity, or the heat continued for sufficient length of time."

Albuminuria is more pronounced in cases in which the alteration in the circulation is due to venous obstruction; especially so the nearer the renal organs the obstruction is situated, as in thrombosis of the inferior vena cava, or the renal veins.

When we take into consideration how many are the causes that temporarily affect the circulation in apparently healthy people, the wonder is, not that albuminuria is occasionally observed in apparent health, but rather that it is not vastly more common. It also appears clear how circumstances, such as chills, overheating, debility, strong mental impressions, use of stimulants, etc., which alter the rapidity or volume of the arterial or venous current, may bring about temporary albuminuria. We perceive, therefore, that albuminuria is a symptom the gravity of which depends altogether upon the cause that induces it, and hence it may mean comparatively little as a symptom, so far as the kidneys are concerned, or it may be one of the utmost gravity, as will be hereafter shown.

As to the next cause of albuminuria—alterations in the vascular walls, it certainly seems most reasonable to suppose that changes in the integrity of vessels containing the blood may readily permit the escape of the more fluid part at least of their contents. We know, indeed, that the smaller vessels are constantly pervious to fluids, which gain ingress and egress in the processes of assimilation, secretion, and excretion. We know that animal membranes permit of the more or less free passage of fluids through them under conditions of perfect health. But albumin being an amorphous or non-crystallizable substance, possesses feebly diffusible properties, and therefore it does not readily filter through membranes in its native state. The conditions in the vessels of the renal glomeruli as to blood-pressure, composition of the vascular walls, and superimposed epithelium, etc., are so accurately adjusted in the healthy organ, that they permit of the filtration of the watery elements of the urine, and perhaps some of the salts in solution only; albumin being retained completely, or, if not completely, the *leak* is so slight that the quantity which escapes is undiscoverable by our present methods of testing.

I may here refer to the doctrine held by Ludwig, and advocated by Senator (op. cit.), viz., that the urine as it leaves the renal glomeruli is always "feebly albuminous," but that the amount escaping is so slight that ordinarily it is undiscoverable, the greater part or the whole being subsequently absorbed by the epithelium of the convoluted tubes during its passage through the latter. Should future observations confirm this hypothesis, it would render still more clear the occasional appearance of albumin in the urine of apparently healthy people: for it is readily conceivable that the conditions of filtration which normally allow only traces of albumin to pass out from the vessels of the tuft, may be easily disturbed and permit its filtration in quantities readily perceptible, even by means of the coarser tests. The normal adjustment of conditions in the renal glomeruli becomes vastly altered upon the occurrence of changes in the vascular walls which are common in renal diseases; and in addition to filtration of urine, albumin also escapes in greater or less quantity as the result. Inflammation is a very common cause of such changes, and it also gives rise to albuminuria with the greatest uniformity. We know that wherever inflammation exists, it so alters the vascular tunics that not only serum of the blood escapes, but also the blood corpuscles themselves migrate through the walls of the finer vessels. Precisely the same thing occurs in the renal glomeruli in inflammation of the kidneys: serum and corpuscles escape and appear in the urine. The albuminuria of nephritis is large in amount, more so than in most other pathological conditions of the kidneys. In lardaceous disease of these organs also albuminuria is profuse, and in this condition the primary effects of the disease fall chiefly upon the vessels of the renal glomeruli.

The influence of the third cause of albuminuria, viz., degeneration of the renal epithelium, has been generally admitted; but only so far as the epithelium covering the glomerular tufts is concerned. This is doubtless because the glomeruli have hitherto been looked upon as the chief source of albuminuria without considering the possibility of other sources in the renal tract. But, as Senator has pointed out (op. cit.), there is no reason whatever for excluding the epithelium of the convoluted tubes from a share in the production of albuminuria.

I have always held that disturbance of the nutrition, and more especially necrosis of the epithelium lining the convoluted tubes, is a very constant cause of albuminuria. ("Pathology of Bright's Disease," *Chicago Med. Journ. and Examiner*, April, 1883.) We must admit that the blood has free access to the basement membrane of the convoluted tubes, and that the epithelium acts either as a filter, or else is selective, performing the office of specific secretion. Now in either case, as Senator forcibly writes (op. cit.), "when their nutrition and function are disturbed, or when the epithelial cells of the uriniferous tubes are in a state of complete decay, albumin will escape from the blood and lymph. . . . Observations which are alleged to prove the contrary must be based upon error or defective investigation, for if not, all our doctrines with regard to specific glandular secretion must be thrown to the winds."

"Neither is there any just reason for assuming that the amount of albumin escaping through degeneration of the tubular epithelium is insignificant. It is probably true that the amount of albumin escaping through the glomerular vessels is much larger in proportion, for in that location there is less protection to the vascular walls; but if albumin escapes at all through loss of tubular epithelium, it is more likely to be considerable, for the tubular tract is so much more extensive compared with that of the glomerular." Lastly, post-mortem examinations have shown that extensive destruction of the tubular epithelium is usual in cases in which albuminuria has been a marked symptom.

The fourth method of production of albuminuria, viz., through alteration in the composition of the blood, is less clearly understood than those just considered; indeed, some reject this as a cause of albuminuria altogether. Recent observations, however, tend to the conviction that more importance must in future be assigned to the hæmatogenous causes of albuminuria.

There remains no doubt, since the experiments of Stockvis and Lehmann, that certain albumins, such as that of egg, if taken in large quantities, are excreted in greater or less amount by the kidneys. These experiments have been repeated and confirmed by Brown-Séquard and others; indeed, there is reason to believe that any albumin entering the blood, if untransformed, acts as a foreign substance, and is excreted by the kidneys. It is probable that in this way occurs the "albuminuria of digestion," or, more properly speaking, of indigestion, following a hearty meal.

The first view held with regard to the production of albuminuria by its discoverer, Cotugno, seems to have been that it was due to some change or modification in the character of the albumin itself in the system, in consequence of which it escaped through the kidneys.

Prout held the same view, and divided albuminuria into two classes, viz., anæmotrophic and hæmotrophic.[1]

Stockvis instituted some experiments with a view to determine the influence of albumin itself in the production of albuminuria when introduced into the blood from without. These consisted in injecting into the blood or under the skin of twenty-three healthy animals, albuminous urine or blood serum of patients suffering from albuminuria. Albuminuria followed in only two cases, in both of which the albumin injected was taken from a patient, who, while suffering from albuminuria, yet had no renal disease discoverable.

While more extended observations are necessary in order to determine the nature and relations of the hæmatogenous changes which bring about albuminuria, yet the tendency of late is strongly toward the acceptation of these as causes of albuminuria in a considerable number of cases.

It is rarely possible to refer albuminuria to any single cause: for in most cases, indeed, several of the causes just considered are more or less concerned in its production. If we take for example inflammation, we shall always find more or less change in the vascular tunics; usually increased blood-pressure; and sooner or later, if the disease continue long enough, nutrition of the epithelium becomes impaired, and necrosis of the cells follows. Likewise in simple pyrexia, wherein we frequently observe albuminuria, several causes of the latter exist, any single one of which it would be difficult to name as the more active one. The character of the blood suffers some change; the blood-pressure is augmented during the period of exalted temperature; and the epithelium appears swollen when viewed under the microscope.

[1] Prout on Stomach and Renal Diseases, pp. 134-144, London, 1848.

SIGNIFICANCE OF ALBUMINURIA.—As to the gravity of albuminuria as a symptom ; the more persistent it is, the more permanent and fixed may be regarded its cause ; at the same time, the opposite does not imply the cause to be either transient or trivial. It must be regarded as exceptional for albuminuria to persist day after day, for any extended length of time, without changes in the kidneys being present. It is true that if the abnormality be measured by the eye, no appreciable change can be made out in exceptional cases; but it must be evident from what has been stated relative to the causes of albuminuria, that some of these are of such a nature as to render them incapable of being recognized by the eye even when assisted by the microscope. It will, I believe, be safer on the whole to look upon persistent albuminuria as morbid, if not pathological in all cases. I have just stated that the causative factor in intermittent albuminuria is not always either transient or trivial. Indeed, I desire here to emphasize the fact that intermittent albuminuria is common in the most grave of all forms of renal disease, viz., renal cirrhosis. In such cases albumin may be present in the urine only at irregular intervals, as after meals or exercise. It may appear on one day, and be absent the next, or for several days or even weeks, and yet the renal lesions may not only be permanent, but incurable, as will be hereafter shown.

The important lesson may be learned from this, that it is necessary after having once discovered albumin in the urine, to search for it over an extended length of time before pronouncing it insignificant in any given case.

It will usually be safe to assume, if the albumin excreted by the kidneys is large in amount, that the lesions on which it depends are extensive, and consequently grave in character; although the gravity is not always in direct proportion to the quantity of albumin lost by the renal organs. It would also be altogether unsafe to assume, when albuminuria is slight, that the cause is not serious, for as in intermittent albuminuria, so when albuminuria is slight, the most serious lesions of the kidneys may be and often are in progress.

Aside then from persistence or large amount of albumin in the urine, it is impossible to predict with any degree of certainty, from the character of the albuminuria, its relative degree of gravity. Its causes, effects, and accompanying symptoms must be looked to for evidence as to the gravity of each individual case, and these will be fully considered in the succeeding chapters under appropriate heads.

RENAL CASTS.—These are so intimately associated with the subject of albuminuria, and, moreover, they so frequently bear directly upon the significance and causation of the latter, that it may not be improper to consider briefly the subject of renal casts in this chapter.

Renal casts are formed in the convoluted, and sometimes in the straight tubes of the kidneys, chiefly associated with pathological changes of those organs; although exceptionally they are said to have been found when no lesions of the kidneys could be made out. As a rule, they accompany albuminuria; in fact, it is doubtful if they ever appear in the urine unless albumin is present, or has recently been present, as will appear from what follows.

Henle was the first to claim, in 1844, that renal casts are composed of fibrin, which has escaped from the blood into the renal tubules, and there coagulated into cakes or moulds; and since then they have generally received the name of fibrinous casts. Recent investigations tend to show that renal casts are composed of albumin, which escapes from the blood or the cellular elements lining the renal tubules. If there be no degenerative changes going on in the tubular elements when the exudation occurs, the casts will be composed of the exudate only, and hence will be clear or hyaline. If degeneration be in progress, the exudate becomes mingled with the detached or degenerated cellular elements, and forms a compound cast; thus epithelial, bloody, fatty, or granular casts may be found, according to the nature and stage of the pathological changes going on at the time and place in which they are formed.

It will be seen, therefore, that renal casts afford valuable information as to the nature of the morbid changes taking place in the renal tubules; and it has been truly said that "they carry with them the very elements of these changes, and present them to our vision when viewed under the microscope."

Renal casts vary in size from $\frac{1}{3500}$ to $\frac{1}{500}$ of an inch in diameter, according to the size and condition of the tubule from which they spring. From what has been said in reference to exudation of albuminous matter into the tubules, it will appear obvious how hyaline or clear casts may be found in nearly all forms of renal disease in which albuminuria is present; their presence alone signifies that no desquamation is going on in the renal tubuli. Epithelial casts consist of the epithelium thrown off from the lining of the tubuli. This may solidify with albuminous matter into moulds; or, primarily hyaline casts may be washed down and carry on their surface more or less detached epithelium from the tubal lining. In whichever way they may arise, epithelial casts indicate the presence of acute inflammation in the tubular structure of the kidney, usually of recent origin. The epithelium may be undergoing fatty degeneration, in which case the fatty elements become entangled in or attached to the casts; these casts indicate subacute or chronic changes. The epithelium may be undergoing necrotic changes and breaking down into finely granular elements, and these may similarly form into moulds of the tubuli, termed granular casts. They indicate more advanced or chronic renal changes. Blood casts are formed by the escape of corpuscles and liquor sanguinis into the tubuli, where the coagulable part of the blood solidifies, forming moulds which entangle the corpuscles. Blood casts, especially when numerous, indicate acute disease, although it is not uncommon to find scattering blood casts in chronic conditions as well.

Finally, renal casts may be, and indeed most often are, of a mixed kind, containing two or more of the elements just described. The most frequent combination found is epithelial with blood—most common in acute conditions; or granular, with fatty elements, forming casts which are most common in chronic conditions.

Casts are rarely found in alkaline urine, no matter what renal lesions exist. In surgical kidney, so called, or pyo-nephrosis, for instance, the urine rapidly undergoes ammoniacal decomposition, rendering it alkaline, and casts are seldom found therein, although the renal changes are very grave.

TESTING FOR ALBUMIN IN THE URINE.

If any doubts exist as to the presence of albumin in the urine, it has frequently been advised, as the safest course to pursue, to collect the whole twenty-four hours' product, and test a sample of the mixture. Now the only object in saving the whole twenty-four hours' product is to detect albumin in cases in which its presence may be intermittent. But if albuminuria be intermittent in character, and the above course be followed, that portion of the urine containing the albumin—which may be but a small portion relatively of the twenty-four hours' product—becomes mixed with the whole; and if the quantity of albumin present be small, it is thus rendered so dilute that it may altogether escape detection.

If the normal conditions of filtration within the renal glomeruli be but slightly disturbed, albumin may escape only under exalted blood-pressure, which mostly occurs during exercise, as already shown. Therefore, the urine voided at night, or in the morning on rising, is often free from albumin; while that passed during the day may contain albumin in perceptible quantity.

In all doubtful cases, therefore, I prefer for diagnostic purposes the urine passed during the hours of most active exercise; for, if the kidneys be defective, the "leak" is most likely to be discovered under such circumstances.

Having secured a sample of urine for examination, its color should be noted, which normally ranges from pale yellow to amber. If the urine be turbid or smoky, it should be filtered. The specific gravity should next be recorded, which normally is near 1020 if taken at a temperature of 60° F. The reaction should then be recorded, and special note taken thereof, as the condition of acidity or alkalinity bears most importantly upon albumin testing, as will be presently seen.

Before dealing with individual tests for albumin in the urine, it may be well to point out the more common sources of error likely to be encountered; as certain bodies more or less frequently present in the urine may cause a precipitate under certain circumstances with the agents employed for detecting albumin and may be taken for the latter unless the observer be on his guard.

URATES.—These salts are feebly soluble, especially at a low temperature; hence, when in excess—as may be inferred if the specific gravity of the urine be high— they are precipitated as the urine cools. If then the urine be of high specific gravity, we should be on the lookout for urates, which will often render the urine turbid before any reagent is added to it. Urates are feebly combined chemical compounds, and are readily separated by a number of agents used in searching for albumin in the urine. Nitric and picric acids, potassio-mercuric iodide and sodium tungstate solutions, and rarely ferrocyanide of potassium solution, cause a turbidity with urates which may be mistaken for albumin. Urates, however, are at once dissolved by heat, which reaction distinguishes them from albumin.

PHOSPHATES.—These salts are maintained in solution in the urine by means of its acidity, and if from any cause the urine becomes diminished in its acidity, neutral, or especially alkaline, they are likely to fall out of solution, and form a precipitate that may be mistaken for albumin. Heat does not dissolve phosphatic turbidity, but on the contrary increases it; and this circumstance increases the liability to error from this source. Acids quickly dissolve phosphatic deposits, which readily distinguishes them from albumin.

MUCIN.—This substance is a derivative of the proteid group, and exists in minute quantities in normal urine. Any irritation of the urinary tract causes its production in relatively large quantities, and it then becomes a source of embarrassment in testing for albumin. The mucin reaction is a common source of error with most of the newer tests, such as picric acid, etc., and in fact mucin is liable to be precipitated by all tests containing acids ; moreover, the precipitate thus formed is not cleared up by heat. Dilute mineral acids, added slowly, precipitate mucin, but redissolve it if added in excess. Organic acids precipitate mucin in urine, but do not redissolve it when added in excess.

PEPTONES.—These, as already said, belong to the proteid group, and constitute a transition state of albuminoids. They occur in the urine occasionally, and when present they may be precipitated by picric acid, and by potassio-mercuric iodide solution "in the cold." They are dissolved by heat, which distinguishes them from serum albumin, the turbidity returning as the urine cools. With an alkaline solution of cupric sulphate, as Fehling's test solution, they produce a rosy-red color (Ralfe).

FOREIGN SUBSTANCES IN URINE.—Certain medicines if taken by the patient, notably some of the vegetable alkaloids, as quinine, strychnia, etc., also some of the oleo-resins as copaiba, appear in the urine, and are thrown down by certain of the tests for albumin, forming a precipitate which may be mistaken for the latter. The picric and mercuric tests are the most likely to cause precipitates with these bodies, although nitric acid "in the cold" may do the same. In all cases the turbidity thus formed is readily dissipated by heat, which distinguishes it from the albuminous reaction.

HEAT.—Boiling the urine was the first test employed for the purpose of detecting albumin ; and for a long time it was the only one resorted to. We now have several tests, some of which are claimed to be more sensitive in detecting the presence of minute quantities of albumin. The great value of heat as a test is as a discriminator of serum albumin from other proteids, rather than as a detector of albumin itself; and until some test shall be discovered which can take its place, it must remain as the essential one to which to appeal when in doubt between serum albumin and other proteids. Heat causes a precipitate in the urine with serum albumin only, saving sometimes with phosphates; but, on the other hand, albumin is capable of combining with an acid or an alkali, in either case forming a compound which is unaffected by heat. This may be readily illustrated by adding a few drops of liquor potassæ to a sample of albuminous urine in a test-tube, when it will be found that boiling fails to precipitate the albumin. Again, if two or three drops of nitric acid be added to a similar sample of albuminous urine, a light cloud will first form, which disappears upon shaking; and if next the urine be raised to the boiling point, the albumin is not precipitated It will therefore be seen that *heat alone* is unreliable as a test for albumin in the urine, as the albumin may exist in an acid or alkaline state; in either case heat will completely fail to reveal its presence.

But if the urine be alkaline, another source of error may be encountered: the phosphates being liable to be precipitated, as already shown, the addition of heat only serves to increase phosphatic turbidity. Such are the errors likely to be met with in testing for albumin with the aid of heat, and the best way to avoid them is to proceed as follows. Pour into a test-tube a column of urine to the depth of three inches, and gradually heat to the boiling point, first having noted the reaction of the urine. The precipitation of urates, peptones, alkaloids, and, in short, all sources of turbidity are thus avoided, save those produced by phosphates and albumin. If the normal acidity of the urine be diminished, next add acetic acid drop by drop, until three or four drops have been added or a precipitate formed, continuing the heating for a few seconds before the addition of successive drops of acid. If albumin be present, heat at or just below the boiling point, combined with some of the above degrees of acidity, will bring it to light with great certainty. If the urine be already strongly acid, as indicated by litmus paper, heat alone will reveal the albumin without addition of acid. If further confirmation be desired, Dr. Tyson directs (*Practical Examination of Urine*, 1886): to the boiling urine quickly add half its bulk of strong potash solution (1 part to 2 of distilled water), when the albumin will be dissolved.

Acetic acid is preferable to nitric acid in testing for albumin with heat, because, on the one hand, it is less likely to overstep the proper degree of acidity, and thus form the acid modification of albumin which is not thrown down by heat; and, on the other hand, it will not, when used as directed, dissolve small traces of albumin which may be present, as may happen with nitric acid.

In order to show this test in the strongest contrast, only the upper stratum of urine in the test-tube should be heated, leaving the lower stratum unchanged. This method is very convenient for comparative purposes.

NITRIC ACID.—This test is best applied according to Heller's method—*i. e.*, by the contact method "in the cold." Clear, chemically pure nitric acid is poured into a test-tube to the depth of an inch. Upon this column of acid a stream of urine is gently floated to the depth of an inch, care being taken to avoid mixing the acid and urine. The urine being the lighter of the two fluids, floats upon the surface, unless propelled forcibly down the tube. To avoid the latter, the stream may be delivered from a nipple pipette, the test-tube being held inclined, which checks the force of the current. When the stratum of nitric acid and that of urine come in contact, if albumin be present in the urine a white wavering zone immediately appears at the junction of the two fluids.

Another method sometimes practised is to reverse the order of introducing the fluids into the tube, the urine being introduced first. In this way the acid, owing to its greater specific gravity, passes through the column of urine to the bottom of the tube. As the two fluids ultimately come to occupy the same relative positions in the tube as in the first method, and moreover as the two fluids are more likely to mix in the last, I prefer the method first described.

This test is one of great delicacy, and, on the whole, very reliable and convenient for office work It has at least two sources of error which should be borne in mind, viz., it may cause a zone of precipitation with urates and with oleo-resins In the former case the zone is less sharply defined and gradually shades into the urine above. The zone of precipitated urates is situated above the line of contact of the two fluids, hence it occurs higher up than is the case with albumin. Lastly, the precipitated urates are dissolved by heat.

The oleo-resins form a yellowish precipitate which disappears on the application of heat. The nitric acid test applied as directed above brings to light all modifications of serum albumin.

PICRIC ACID.—This test solution may be prepared by saturating boiling distilled water with picric acid in powder —about six or seven grains to the ounce—and filtering. Owing to the fact that this test gives a reaction with a number of bodies other than albumin met with in the urine, I was disinclined at first to favor its use. Since, however, I have seen it applied in the hands of its discoverer, Dr. George Johnson, who kindly explained to me his method, I am satisfied that in picric acid we have a test of such great delicacy that we cannot afford to dispense with it; moreover, with proper precautions it is one of reasonable reliability.

For the purpose of detecting minute traces of albumin in urine Dr. Johnson proceeds as follows : Into an ordinary sized test-tube is poured a column of urine to the depth of four inches; then holding the tube inclined, an inch of picric acid solution is gently poured on the surface of the urine, which mixes with only the upper layer of the latter; and as far as the yellow color of the picric solution extends, the coagulated albumin renders the urine turbid, thus contrasting with the transparent, unstained urine below. The application of heat increases the turbidity, due to precipitation of albumin. If the tube be then set aside for an hour, the coagulated albumin subsides, forming a delicate film at the junction of the colored and the unstained strata of urine. For the action of this test there must be an *actual mixture*, and not a mere surface contact of the two liquids.

It is true that picric acid precipitates peptones, alkaloids, oleo-resins, and rarely also urates; but all of these are readily cleared up by heat, which distinguishes them from albumin.

The great source of error with picric acid, as with several other tests for albumin, is the liability to throw down mucin, which may be taken for albumin, as the former is not dissolved by heat.

POTASSIO-MERCURIC IODIDE.—This test solution is prepared by mixing 2.7 parts of mercuric chloride and 6.64 parts of potassium iodide with 100 parts of water.

I prefer to apply this test as follows: An ordinary test-tube is filled with the suspected urine to within an inch of the top. The urine is next acidified by the addition of a few drops of acetic acid, or saturated solution of citric acid. The tube is allowed to stand a minute or two, when, if mucin be present in the urine to any extent, a slight turbidity will be apparent. *One-half* of the mixture is next poured into a similar sized test-tube, and a few drops of the mercuric solution are added to the remainder, when the two tubes are compared side by side. If the turbidity be increased by the addition of the mercuric solution, and be not reduced by heat to the degree of opacity in the second tube, albumin is present. Or, if after the addition of the acid to the urine no change takes place after standing for a minute or two, the mercuric solution may be added to the *whole quantity* of urine in the tube, and any turbidity resulting which is not dissipated by heat is due to albumin.

For the action of this test the urine requires to be acidified, preferably by acetic or citric acid.

This test is probably the most sensitive of all albumin precipitants. Its sources of error are that it precipitates peptones, urates. alkaloids, or oleo-resins; but all of these are cleared up by heat.

In testing, mucin also may be thrown down, which is not cleared up by heat; but if the method of testing be conducted as directed, this source of error will be discovered before the reagent is added.

FERROCYANIDE OF POTASSIUM.—This test solution is prepared by simply saturating cold distilled water with ferrocyanide of potassium. The urine to be tested must first be acidulated by the addition of a few drops of acetic acid, or saturated solution of citric acid, after which, upon the addition of a few drops of the ferrocyanide solution any albumin present will be thrown out of solution, causing a white turbidity. Some time since I called attention to the fact that this test much less frequently produces a precipitate with other bodies besides albumin than most other tests. (*Journ. Amer. Med. Assoc.*, Jan. 19, 1884.) Subsequent use has shown me that this test is the least liable to errors of all tests for albumin in the urine if properly applied.

Ferrocyanide of potassium solution *does not* precipitate peptones, alkaloids, phosphates, or oleo-resins, in urine; but it precipitates all modifications of serum albumin. Rarely, if the urine be of high specific gravity, this test may precipitate urates; but this can be avoided by diluting all urines of high gravity to the normal standard (1020), or slightly below, by the addition of distilled water, before testing. But even without the precaution of reducing the specific gravity of abnormally heavy urines, the precipitation of urates by the ferrocyanic test is an extremely rare occurrence, as shown by the fact that it has occurred in my hands only half a dozen times during the almost daily use of the test for the past seven years.

The reaction of the ferrocyanic test with albumin is somewhat slower in appearing than is that of some of the other tests. Its range of sensitiveness in detecting albumin is about the same as that of the nitric acid test (Heller's).

Other tests have been recommended for the purpose of detecting albumin in the urine, among which may be mentioned acidulated brine, proposed by Dr. Roberts (*London Lancet*, October 14, 1882), and also the sodium tungstate test introduced by Dr. George Oliver (*London Lancet*, February 3, 1883). As these possess no special merits over those already described, and more especially since their use has been discontinued by their discoverers, their description is omitted.

Of the five tests for albumin in the urine that have

just been considered, the most sensitive are the mercuric and picric; next in order are heat, ferrocyanic, and nitric acid in the cold. The most reliable are the ferrocyanic and nitric acid (Heller's). It may be thought unnecessary to advise the use of so many tests for albumin; some have claimed that one or two are quite sufficient. I know of no test for albumin in the urine which, under some circumstances, has not misled the observer; and it is therefore my habit, in those rarer cases in which doubt arises, to submit the urine in succession to each of the five tests described. Under circumstances of doubt the five tests are none too many to settle the uncertainty. The crudest test will scarcely fail to reveal the presence of albumin if present in large quantities; but in cases in which only slight traces are present, it is altogether different; in such case nitric acid, for instance, may fail to reveal them. Now it is these smaller quantities of albumin in the urine that are frequently of the gravest clinical significance, as has already been shown in the case of renal cirrhosis. Again, in scarlatinal nephritis, slight traces of albumin may be overlooked if we rely on coarse tests, and the patient may be dismissed as cured while, at the same time, more or less mischief is in progress in the kidneys, and after a time the patient may return with hopelessly damaged kidneys. It is impossible to insist too strongly upon the necessity of guarding against such errors by the most searching examination of the urine. An impression seems to have gained considerable belief with the profession that the smaller traces of albumin in the urine are of little clinical importance; but it would be difficult to find a more signal mistake in the whole range of clinical medicine, or one behind which more danger to life lurks. It is, therefore, with the view of avoiding such errors, that the following systematic course of proceeding is advised.

In testing for albumin in the urine, for the first time, always employ at least two tests. These should consist of one of the most sensitive, as the mercuric, and one of the most trustworthy, as the ferrocyanic. In case no reaction be observed with either, it may be safely assumed that no albumin is present. If, on the contrary, a reaction be obtained with one, or both, then the whole five should be systematically gone through with, as already described; and nothing short of this methodical appeal to our resources can excuse any error in this important matter.

In addition to the tests which have been described, Dr. George Oliver (op. cit.) has introduced tests in paper form for detecting both albumin and sugar in the urine. These are extremely convenient, as they may be carried in the pocket, and applied at the bedside of the patient. I have used these papers constantly since their introduction, and find them especially valuable in clinical work, where it is so often desirable to ascertain the condition of the urine at the same time that the general symptoms of the patient are passing under inspection. Nearly two years ago, Messrs. Parke, Davis & Co., of Detroit, kindly prepared, at my request, the whole series of these test-papers, so that they are now obtainable in this country.

To Messrs. Parke, Davis & Co. is due especial credit for arranging the tests in the form of a very neat pocket-case in the smallest possible compass. By means of these tests a comprehensive knowledge of the characters of the urine may be readily ascertained at the bedside of the patient.

The most reliable tests in paper form for albumin are the potassio-mercuric iodide, and the ferrocyanide of potassium. They should be used in conjunction with a citric acid paper in all cases.

Dr. Oliver's[1] method of applying these tests is as follows: A mercuric or ferrocyanic and a citric acid paper are introduced into a small test-tube containing sixty minims of water. After gentle agitation for half a minute or so, the test-papers are removed, and the trans-

FIG. I.

Pocket-case.

parent solution is ready for testing. The pipette containing the suspected urine is held vertically over the tube, and the urine is delivered in drops; any reaction which occurs is readily observed by holding the tube against a dark background in a good light. In the case of the mercuric test, if a reaction occur, heat must be applied

[1] Bedside Urine Testing, 1885, third edition.

in order to clear up any opacity not due to albumin. The use of heat is unnecessary in testing with the ferrocyanic paper; indeed, this test applied as directed is remarkably free from errors and may be confidently relied upon. The paper tests for albumin should never be introduced into the urine direct, for if albumin be present in large quantity it may coagulate upon the reagent paper and prevent the latter delivering its charge of the reagent to the urine; in which case no reaction in the urine will be perceptible.

The quantity of albumin in a given specimen of urine may be determined by coagulating it by means of heat, or ferrocyanide of potassium and acetic acid, and after filtering and drying it may be weighed. This process involves the expenditure of considerable time, and the information gained thereby is scarcely of sufficient clinical importance to call for its use in routine work. As has been shown, the gravity of a given case of albuminuria by no means depends necessarily upon the quantity of albumin present in the urine. It is not the absolute quantity of albumin excreted that it is so desirable to determine, but the relative quantity passed from day to day; for it is popularly believed, and the belief is probably well founded in most cases, that the course of the renal lesion keeps pace with the *relative* amount of albumin excreted by the kidneys from day to day. In observing the effects of treatment, therefore, it is more desirable to know whether the quantity of albumin is increasing or decreasing, than to know the exact quantity present in the urine.

A useful method of determining the relative excretion of albumin by the kidneys, especially in cases in which the amount is considerable, is the following, which requires but a trivial expenditure of time and labor:

Fill a graduated test-tube with the albuminous urine, add a few drops of acetic acid and ferrocyanide of potassium solution: all of the albumin is precipitated, and slowly settles to the bottom of the glass. Set the tube aside for twenty-four hours, when the quantity of albumin by bulk may be read off from the scale on the side of the tube. On comparing this test from day to day, approximate results may be obtained which are fairly satisfactory for practical purposes. It should be borne in mind that with the above, or with any other quantitative or volumetric method of estimating albumin, observations must be conducted upon a part of a mixture of the whole twenty-four hours quantity of urine, in order to be of definite value.

If more accurate results be desired, or if the quantity of albumin present be very small, too small to be measured by the above process, then the quantitative method recently suggested by Dr. Oliver (op. cit.)—to whom we owe so much for time-saving methods of urinalysis—will be found the most convenient. Dr. Oliver's method consists in precipitating the albumin from a given quantity of urine by a mercuric test-paper, and comparing the resulting opacity with that produced by coagulating the albumin from a standard solution containing one-tenth of one per cent. of albumin. This standard of opacity is imitated by a piece of ground-glass. The test is applied as follows: The suspected urine is first diluted by the addition of twice its bulk of water. Into a flattened test-tube, graduated to 200 minims, in 10 minim divisions, are poured 50 minims of the diluted urine, and a potassio-mercuric and a citric acid paper are dropped into the tube. The tube is well shaken for a minute or so, and the test-papers withdrawn. The resulting turbidity is measured by placing behind the tube a card on which lines of different degrees of thickness are printed. If the opacity be such as to obscure the lines completely, water is added slowly until the opacity of the solution corresponds exactly with the ground-glass standard. A comparison between the known value of the precipitant test-paper, and the number of times the urine has been diluted, furnishes the proportion of albumin. The value of the mercuric papers is one-tenth per cent. Thus, for example, if 50 minims of water are required to reduce the opacity of the diluted urine to the exact standard of the ground-glass, the solution contains two-tenths of one per cent. of albumin. But as the urine was first diluted with double its volume of water, the above result must be multiplied by 3, which gives six-tenths of one per cent., the actual percentage of albumin present.

HYGIENIC TREATMENT OF ALBUMINURIA.

Certain general rules, chiefly of a hygienic character, are more or less applicable to the management of all cases of albuminuria; and with a view to avoid frequent repetitions of these in succeeding chapters, they will now be considered.

DIET.—First in importance may be ranked the diet of the albuminuric patient. It was formerly held that the loss of albumin by the kidneys, more especially in cases in which albuminuria is marked, represents a very serious tissue-waste which should be made up by a liberal use of such foods as furnish to the system the largest proportion of these elements. But that such doctrine is not only not tenable, but a very dangerous one to carry into practice in many cases of albuminuria, can no longer be doubted. In the first place, it is a fact beyond dispute that the quantity of albumin contained in an ordinary meal upon meat diet exceeds the loss sustained by the system for nearly a week; not only in ordinary cases, but even in those in which albuminuria is considered to be excessive. We must look not alone to the loss of albumin for the explanation of the defective nutrition so common in Bright's disease; but more especially to the loss of corpuscles and surcharging of the blood with effete products which render that fluid unsuitable for the healthy nourishment of the organism. It must appear clear that the use of albuminous foods, the decomposition products of which furnish such large quantities of nitrogenous waste, not only aggravates the evils which it is intended to remedy; but it also invites danger from defective elimination, and consequent accumulation in the system, favoring the production of uræmia.

Again, albuminous food acts injuriously upon the kidneys themselves. Lehmann and Stokvis have shown by experimental investigations that when egg albumin is introduced into the blood, not only is it excreted by the kidneys, as was first demonstrated by Claude Bernard, but serum albumin of the blood also makes its escape with it; in other words, more albumin escapes from the system under such circumstances than is introduced. Therefore, some forms of albumin, at least, act as irritants to the kidneys when excreted by those organs. It is found that egg albumin, when introduced into the system, even by way of the stomach, especially in liberal quantities, also excites albuminuria quite frequently in apparently healthy people. Christison has pointed out that the same result follows the use of cheese; and, moreover, temporary albuminuria, as already shown, is by no means a rare consequence of the ingestion of large quantities of proteid foods.

If, therefore, we accept the proposition which has been conclusively proved by Lehmann and Stokvis, viz., that the excretion of albumin by the kidneys induces irritation of those organs, then we must conclude that the liberal use of albuminous foods is injurious to albuminuric patients; for, in such cases, a part at least of the albumin is eliminated by the kidneys.

We know now that the requirements of the economy do not demand as large quantities of nitrogenous foods as was formerly supposed. Recent investigations seem to establish that fats, at least to some extent, are the ultimate products of the metamorphosis of nitrogenous foods in the system, as shown by Lawes and Gilbert ("Sources of Fat of the Animal Body," *Philosophical Magazine*, vol. xxxii.); consequently, we may substitute the fats largely for nitrogenous food in cases of albuminuria, thus more quickly reaching the proximate elements of nutrition, and at the same time avoid clogging the system with unnecessarily large quantities of nitrogen, which latter is the principal element of waste eliminated by the kidneys. As legitimate deductions from the foregoing, the following rules as to diet are given as applicable in cases of albuminuria:

Lean meats should be more or less rigidly excluded from the diet, according to the acuteness or otherwise of the symptoms. In very acute cases, no animal food whatever, not even soups, should be used. In subacute and chronic cases, meat may be permitted once a day, or two or three times a week. Fish is the least objectionable; the white meat of fowl, as chicken or turkey, next. Young and tender beef or mutton may be permitted in some cases; but all seasoned or cured meats, as ham, tongue, sausages, hash, corned beef, etc., should be avoided in all cases. Fat meats may be permitted as freely as the stomach is able to digest them. Eggs and cheese must not be used. Milk is a good article of diet, as it contains only about four and a half per cent. of nitrogen. In all save acute cases, milk may be used freely if it agree with the patient. In some it will be found to suit excellently, and in such cases the effects of an absolute milk diet are often strikingly beneficial. •In most cases milk agrees best when skimmed. Boiling usually renders it more acceptable to the stomach. In some cases milk in any form, will be found to disagree, and in such it is better to abandon its use and seek some substitute. I have seen albuminuria alarmingly increase by endeavoring to reconcile the system to the use of milk.

Farinaceous articles and fruits may be taken *ad libitum*, as may all vegetables, except leguminous ones. Finally, as already pointed out, overloading the stomach often causes slight albuminuria, and from this we are taught the important lesson that food should be partaken of in moderate quantities at a time.

Great discrimination is requisite in the use of beverages in albuminuria; above all, in the use of alcohol. It cannot be longer denied successfully that alcohol in any and all forms is injurious to kidneys which are excreting albumin. It has long been held that alcohol is one of the frequent causes of Bright's disease, especially when used in a concentrated form. I must, from my own observations, declare my concurrence in this belief. A very large proportion of my cases of albuminuria have occurred among spirit-drinkers, and I have so often observed an outbreak of albuminuria as a sequel to an unusual indulgence in spirits, that I believe alcohol to be a not infrequent cause of renal inflammation. But it is no longer necessary to appeal to personal experiences in this matter, for Penzoldt has proved by experiment (*Litzgsb. d. phys. med. Soc. zu. Erlangen*, Juni 19, 1882) that alcohol, if injected into the blood of animals, causes both albuminuria and nephritis. We must, therefore, forbid the use of alcohol by our patients if we undertake conscientiously to place them under the most favorable conditions for recovery. There are cases, it is true, in which long-continued use renders absolute abstinence difficult; but as this indulgence may be the exciting cause of the albuminuria, there is the greater necessity for its discontinuance. The life of the patient is altogether too valuable to be placed in the balance against a mere question of habit.

The liberal use of fluids between meals should be encouraged in all forms of albuminuria. They act as the simplest and least stimulating of all diuretics by increasing the volume of the blood in the vessels, thus raising the tension in the latter, and facilitating urinary filtration. They also dilute the waste products in the blood, and holding these in solution as they pass out, exert a depurative action on the organism. It is in this way that mineral waters often prove highly beneficial in albuminuria; those which are the purest—*i. e.*, freest from organic matter—are the most desirable. Ordinary distilled water is a desirable beverage in these cases, as its freedom from foreign matters enables it to absorb more readily the waste products in the organism and carry them in solution from the system.

THE SKIN.—It is of the greatest importance to secure uniform activity of the cutaneous function in all forms of albuminuria. The intimate functional relationship existing between the skin and kidneys has long been known, the former being capable, to some extent, of acting vicariously to the latter; thus compensating, in no small degree, for defective renal action. For the purpose of securing the fullest benefits to be derived from activity of the cutaneous function, various baths are of the greatest service, and these should always be used warm. The hot-water bath, vapor, pack, or, best of all, dry hot air, produce excellent results, not only in eliminating waste products, thereby lessening the work of the kidneys, but also by inviting the blood to the surface, thus relieving renal congestion. The effect of the warm water bath is heightened by the addition of salt thereto; for, as Senator has shown (op. cit.), the presence of the saline keeps up, for some time, slight irritation of the skin, thus maintaining cutaneous hyperæmia.

Cold in any form applied to the skin is injurious in albuminuria; acting as a powerful vaso-constrictor to the cutaneous vessels, it not only checks transpiration, thus throwing additional work upon the kidneys; but it also markedly increases renal congestion, by diminishing the volume of the blood in the cutaneous capillaries, which goes to over-fill the visceral circulation.

Cold baths are therefore highly pernicious to albuminuric patients, and must be strictly prohibited. The skin should be kept carefully guarded against chilling atmospheric currents for the same reason; and this is best secured by wearing woollen garments next the body during the whole year. The most thorough protection to the skin is secured by the use of the Jaros Hygienic Wear, especially during the cold and damp season in northerly climates. This wear consists of an external cotton fabric, moderately fine in fibre, into which is woven soft fleece wool, the latter to be worn next the skin. It will be seen that the cotton fabric secures a comparatively static condition of air next the skin, while the wool combines the minimum radiation of heat from the body with the greatest attainable hygroscopic power, thus securing uniform dryness and warmth of the skin under varying conditions of atmosphere. Dr. L. L. McArthur has demonstrated before the Chicago Medical Society (*Chicago Med. Journ. and Examiner*, Feb. 1886) the superiority of the Jaros Wear over all other fabrics for the purpose of protecting the skin against rapid changes of atmospheric temperature and humidity, and my own experience with it altogether bears out his deductions.[1] The feet should be warmly clad, even though the patient keep indoors. The great benefit derived from securing uniform warmth and dryness of the skin in cases of albuminuria is shown by the fact that confinement to bed for a time often produces marked improvement.

EXERCISE.—This should always be limited in cases of albuminuria. It is a well-established fact that active muscular exercise increases albuminuria, the reasons for which have been fully explained. Passive exercise only should therefore be indulged in by albuminuric patients; fatigue is to be avoided under all circumstances. It has often been observed, too, that uræmia is especially prone to occur during or after some unusual muscular effort. At first this was ascribed to increased liberation of urea by the muscular contractions. As yet, there is no direct proof that urea in the circulation is the cause of uræmia, although it is probably one of the elements of the cause; and, moreover, it is doubtful in the light of recent investigations if muscular exercise increases directly the amount of urea in the blood, and, therefore, we must await some future explanation of the foregoing fact.

Female patients the subjects of albuminuria should be confined to bed during the menstrual period. I have repeatedly observed the fact pointed out by Senator, viz., "that the albumin is regularly increased in female patients during menstruation." The pelvic congestion consequent upon menstruation doubtless in part, at least, explains this temporary increase of albumin. Sudden rises in the general blood-pressure should, therefore, be guarded against during menstruation, and the most certain way to accomplish this is by enforcing muscular quietude.

[1] The Jaros Wear may be obtained at 103 State Street, Chicago.

PSYCHICAL INFLUENCES.—The opinion has been gaining ground, of late, that mental influences have much to do with the causation of some forms of renal disease. Observers have so frequently noted their development in those subjected to severe mental strain, that, while the exact causative relationship is not as yet understood, it is no longer reasonable to doubt its existence. That the kidneys are susceptible of strong vasomotor influences as the result of central impressions is a matter of daily observation; as examples, may be mentioned the profuse diuresis consequent upon hysteria or fright. It is not improbable that powerful and prolonged cerebral impressions may influence the renal organs in a more permanent and serious manner. At any rate, it is certain that the more powerful emotions often increase the excretion of albumin in those who are the subjects of albuminuria; and, therefore, in such cases it is advisable to maintain the mental condition of the patient in as tranquil a state as possible.

CLIMATE.—This has a most important influence upon albuminuria, and it should in all cases, where practicable, be taken advantage of in the treatment. Dr. Dickinson, to whom we are greatly indebted for investigations in this matter, in referring to the most serious of all forms of albuminuria, writes : . . . " the advantage to be expected from a change of sky is at least as great in renal as in pulmonary disease. Cure is a word to be used with caution, but I have seen little less; the albumin reduced to a trace, and perhaps that inconstant, and the general health brought up almost to its original level."

The most desirable conditions are an equable, dry, and warm climate, with a mean range of temperature as near 65° F. as possible. On our own continent, the southern interior offers especial advantages. Thus, the town of Saltillo, the capital of the State of Coahuila in Northern Mexico, is probably unsurpassed in equability of its climate; observations extending over a number of years show that the temperature never rises above 86° F., or sinks below 50° in the daytime during the whole year round. In addition to Mexico, South Carolina, Georgia, and, above all, Florida, afford excellent climates for resort for these patients; the atmosphere combining warmth and equability to a remarkable degree. To those who are most benefited by sea-air, Nassau and Bermuda are especially to be recommended. In the Eastern Hemisphere, the Mediterranean Coast seems to offer the greatest advantages; thus, Rome, Naples, and the South of France are the most suitable resorts in Europe; while Cairo and Algiers are among the most desirable on the Southern Mediterranean Coast. Albuminuric patients should spend at least their winters in some such places as have just been mentioned, if it be practicable for them to do so. Of course, with the poorer class of people the benefits to be derived from change of climate unfortunately are usually beyond reach, and in such cases the greatest safety is secured by taking these patients into a well-conducted hospital during the cold and damp seasons.

Such are the leading hygienic measures to be employed in conjunction with the medicinal treatment of albuminuria; and it is not too much to assert that on their careful observance a successful issue will, in many cases, be more largely due than to medication. Indeed, the hygienic management of albuminuria has not heretofore received the consideration which it deserves; but recently, increased attention has been given to the subject, so that it is not unlikely soon to occupy the more prominent place among the curative measures of albuminuria to which its merits so justly entitle it.

CHAPTER II.

URÆMIA.

IF the integrity of the kidneys suffers sufficient impairment through disease, as greatly to diminish or altogether to abolish their function, sooner or later, if the condition continue, a remarkable train of symptoms follows of exceedingly dangerous and often fatal character. These symptoms have been grouped together under the term uræmia. As the structural diseases of the kidneys which we propose to consider in the succeeding sections may at any stage of their course become complicated by this affection, it may be well to take a practical survey of the subject here, and thus avoid the necessity of future repetition.

ETIOLGY.

Probably in no field of scientific investigation have more numerous and earnest efforts been put forth than in this, with the view of ascertaining the precise cause of these remarkable manifestations; but, beyond the gross fact that the abolition of the urinary secretion entails sooner or later the appearance of those dangerous symptoms yet to be described, we know comparatively little of their cause.

Christison first claimed that uræmia was due to the presence in the blood of large quantities of urea, which accumulated through failure of the kidneys to eliminate it from the system. For some time this doctrine was accepted by most authorities, and, indeed, at the present time it has some supporters. It has been conclusively shown, however, that while the blood may in some cases be found surcharged with urea, yet quite often it is not so; in fact, it may be conspicuously the opposite. Moreover, the injection of large quantities of urea into the blood of animals has not uniformly produced uræmic symptoms.

Frerichs next claimed that uræmic symptoms were due— not to the presence of urea itself—but rather to its decomposition in the blood "under the influence of some special ferment into carbonate of ammonium." As proof of this, he claimed that carbonate of ammonium was present in the breath of those suffering from uræmia, and, moreover, that the injection of carbonate of ammonium into the blood "evoked all those symptoms which we recognize under the term uræmic." For some time this explanation was generally accepted as correct. Subsequent investigations have failed to demonstrate the presence of carbonate of ammonium with any uniformity or frequency in the blood of those who have died from uræmia; or in the blood of animals that have been fed on large quantities of urea, or in those from which the kidneys have been removed. The frequent presence of carbonate of ammonium in the breath of those suffering from uræmia is now accounted for on the more probable theory that urea is transuded on the mucous surfaces, where it becomes decomposed, thus liberating ammonium carbonate; and strength is lent to this explanation by the fact that carbonate of ammonium is frequently present in the breath of those who have never been the subjects of uræmia. But Schottin showed that the injection of sulphate of sodium or potassium into the veins of animals produced symptoms similar to those which resulted from the injection of ammonium carbonate, which is probably the strongest argument against the doctrine of Frerichs.

Trietz claimed that carbonate of ammonium was the active cause of uræmia, but that it was formed in the intestines (not in the blood), and was reabsorbed into the blood, where it produced the uræmic symptoms. As this is merely a modification of the doctrine of Frerichs, the arguments already noted as opposed to the latter apply also to that of Trietz.

Oppler suggested that the symptoms are excited not by the presence of any element of the urinary secretion in the blood, but rather by the derivatives or the primary elements of tissue metamorphosis which go to make up urea, uric acid, etc. His conclusions were drawn from the fact that he found in the blood of uræmic patients large quantities of such extractives as creatin, creatinin, etc., and in the muscles large quantities of leucin.

Voit suggested that the uræmic phenomena were evoked by the presence of potash salts in the blood, and that these resulted from the retrograde metamorphosis of muscular tissue when the renal function became impaired. D'Espine has found an increase of these salts in the blood in uræmia of scarlet fever, and he' attributes the symptoms to their toxic effects (Ralfe). Besides these toxic or chemical causes, various observers have sought to explain uræmic symptoms on pathological grounds. Osborne thought that meningitis was the cause both of the convulsions and coma of the uræmic state. Traube held that œdema of the brain was responsible for the symptoms, and that this was brought about by the hydræmic condition of the blood in albuminuric patients, aided by hypertrophy of the heart, and high vascular pressure. Meningitis and œdema of the brain, however, are by no means constant accompaniments of uræmia; and when present they are usually associated with convulsions, to which they are more likely due. It is now known also that in those cases of nephritis which are most likely to entail active uræmic symptoms, the blood is the least hydræmic, and, moreover, hypertrophy of the heart is least common in such cases.

Such are the outlines of the more prominent views which have from time to time been held as to the causes of uræmia, and which even at the present time are by no means harmonized. It would be out of place here to enter minutely into the various discussions which have arisen pro and con, in reference to each of these. Suffice it to say, that up to date uræmia has successfully defied the fixation of its cause upon any element of the urine, or its derivatives. All exclusive theories have thus far been negatived, and the few facts we possess point strongly to the probability of several urinary elements or their derivatives being concerned in the etiology of this affection. Recent investigations seem to support this view; thus Prof. Bouchard has lately demonstrated the toxicity of normal urine when injected into the blood (though this had previously been denied), the symptoms evoked thereby in animals closely resembling many of the uræmic order (*British Medical Journal*, June 6, 1885).

The retention, then, in the blood of urinary elements and their derivatives (the latter consisting largely of the waste products of retrograde metamorphosis of tissue), either in whole or part, reacts in a damaging manner, not only on the nervous system, which is probably the most sensitive to the toxic influence, but also on other systems, as the respiratory, circulatory, and the digestive. Not only are the elements of the urinary secretion retained within the organism through renal defect, but also healthy tissue metabolism becomes impeded, clogged, and perhaps almost arrested. As Schottin has shown, suppression being usually associated with uræmia, a large amount of acid, which normally passes out through the kidneys, is retained in the system, reducing the alkalinity of the blood, and enfeebling its oxidizing powers.

It has been claimed, also, that the loss of albumin, the drain of corpuscles, and consequent hydræmic conditions of the blood are concerned in the causation of uræmia; but, on the one hand, there is no constant coincidence of these states of the blood with uræmia; and, on the other hand, these conditions never bring about uræmia when resulting from other than real causes.

In the light of what is known, then, the only conclusions that seem justifiable are, that the cause of uræmia lies in the failure of the kidneys to excrete the urine, in part or in whole, or in its primary elements; and that these either act as toxic agents on the system, or generate some new, and, at present, unknown agent (through conditions resulting from their accumulation), which acts as a poison on the organism, evoking the symptoms which we term uræmic.

SYMPTOMS.

It will be most convenient, as well as practical, to consider the symptoms of uræmia under two forms, acute and chronic; depending on intensity and duration of the cause, and consequent symptoms. As in the great majority of cases the less active symptoms prevail for a longer or shorter time, before acute uræmia breaks forth, it will, therefore, be more convenient to consider chronic uræmia first.

CHRONIC URÆMIA.—As impairment or suppression of the renal function is the primary element in the causation of uræmia, so it is the primary symptom of this disturbance,

SYMPTOMS. 63

preceding invariably all others, of which it becomes a valuable index, and if properly interpreted will usually put us on our guard as to its approach. In the management of all renal diseases, then, it is essential to be informed, not only of the actual daily quantity of urine passed, but also of the *average quantity* passed in the course of the disease. My own experience has been that when the average quantity of urine falls off to half the normal amount, chronic uræmic symptoms usually follow sooner or later, if the condition persists. It is not, as a rule, till oliguresis reaches a much lower point, sometimes, indeed, not till complete suppression occurs, that the acute symptoms are manifested. Of course, the degree of diminution necessary to bring about uræmia, acute or chronic, can never be predicted with certainty, because other circumstances more or less modify the result, as the amount of solids contained in the urine, the quality of food taken, the degree of muscular exercise indulged in, and the length of time oliguresis has been present, etc. Notwithstanding all this, there are certain limits, as already indicated, within which more or less danger may be apprehended. It has been claimed that the quantity of urine may be normal, or nearly so, and yet uræmia result, owing to the non-excretion of the solids with the urine. I have never known acute uræmia to result when the *average quantity* of urine has been normal for a considerable length of time, no matter how low its specific gravity. I am aware that the aqueous elements of the urine, for the most part, have their source in one part of the secreting structure of the kidneys, while the solids are mostly contributed from another; and thus their elimination would seem to be independent and separate. As a matter of fact, however, so far as my own observations are concerned, they are not *entirely* independent; and an average normal quantity of urine is incompatible with a diminution of the solids to the extent of producing uræmia. This may not accord with present theories, but it does with facts. Of course, if the quantity of urine has for a long time averaged considerably below normal, till the system has become charged with toxic elements, it is quite conceivable how, even after the tide has turned, uræmia may be precipitated through inordinate exercise, or the ingestion of a large amount of nitrogeneous food, or both; and yet the actual quantity of urine excreted at that particular time might be normal. Those cases of uræmia which are alleged to have occurred when the urine was normal in quantity, I believe must have arisen under the above circumstances. The quantity of solids excreted by the kidneys—circumstances being equal—indicates the functional capacity of those organs. It is important, therefore, to note the amount of the solids contained in the urine in all cases; and for such purposes observations must be conducted upon a part of a mixture of the whole twenty-four hours' product to be of definite value. Now it is obvious that bodily weight, exercise, and amount and quality of food taken, will more or less influence the actual quantity of solids thrown out by the kidneys, and hence these will be found to vary within certain limits. It is known that the amount of solids excreted by the kidneys, under ordinary circumstances, in the average healthy individual is about two ounces in twenty-four hours, or four per cent. of the whole urinary product. It is easy to estimate approximately the quantity of solids excreted if we know the specific gravity, and the whole quantity of urine passed. Several rules for the approximate estimation of solids in the urine have been devised, the most widely known of which are those of Trapp and Haeser. The simplest rule, however, is to multiply the last two figures of the specific gravity by the number of ounces of the urine, and the product will approximatively represent in grains the "solid urine" discharged in twenty-four hours (Oliver). In making the above estimate, the specific gravity of the urine should always be taken at a temperature of 60° F., as urinometers are usually corrected from that point.

The relation between the quantity of urea in the urine and the production of uræmia has heretofore received much attention, and various complicated and time-consuming methods have been devised for estimating accurately the discharge of urea by the kidneys. Now the quantity of urea contained in a given urine comprises about one-half of the whole solid urine under ordinary circumstances. But in diseased states of the kidneys the proportion of urea often suffers a greater comparative reduction than do most of the other solids of the urine; at any rate, trustworthy observers have claimed that in many cases a very notable diminution in the proportion of urea in the urine takes place just prior to the outbreak of acute uræmia. Although it has never been conclusively shown that retained urea is the cause of uræmia, yet for the reasons given, as well as the fact that the quantity of urea discharged furnishes us a gauge of the functional capacity of the kidneys, it is well to be able to estimate readily the quantity of urea contained in the urine.

A very accurate, and certainly the most simple method with which I am acquainted is that devised by Dr. George B. Fowler,[1] of New York, to whom I am indebted for a special explanation of his process, of which the following is a brief synopsis:

[1] Prize Essay, Alumni Association, College of Physicians and Surgeons, New York, 1877. New York Medical Journal, June, 1877.

The principle of the process depends upon the facts, (*a*) that it is the urea, chiefly, which gives the urine its specific gravity; (*b*) that when the urea is decomposed by mixing with the urine a solution of chlorinated soda (Labarraque's solution), the specific gravity of the mixture is diminished; and (*c*) that there is a definite relation between the loss in specific gravity and the proportion of urea originally present.[1]

A decrease of one degree of specific gravity corresponds to $3\frac{1}{2}$ grains of urea per ounce of urine, or 0.77 per cent.

The method consists in adding to one volume of urine (say one ounce), in a large hydrometer jar, seven volumes of Labarraque's solution. Decomposition immediately ensues, and at the expiration of a few hours, the vessel being occasionally shaken in the mean time, all the nitrogen of the contained urea has escaped. The specific gravity of the quiescent mixture is now to be noted, and also that of the pure urine. We now have the specific gravity of the mixture of the urine and Labarraque's solution *after decomposition*. In order to ascertain what it was before decomposition, we must resort to the law of proportions. Multiply, therefore, the specific gravity of the Labarraque solution by 7, add the specific gravity of the urine, and divide the sum by 8. Now from this—the specific gravity of the mixture before decomposition, subtract that obtained after decomposition. Multiply the difference in degrees by $3\frac{1}{2}$, and the result will be the number of grains of urea per ounce of the urine, or, better, by 0.77, which gives the percentage of urea. The presence of

---

[1] The formula for the reaction, by which all the nitrogen is set free, is as follows:

$$\frac{\text{Urea}}{CH_4N_2O} + \frac{\text{Labarraque}}{3NaClO} = \frac{\text{Carbon. acid}}{CO_2} + \frac{\text{Sod. chlor.}}{NaCl} + \frac{\text{Water}}{2(H_2O)} + N_2.$$

sugar or albumin in the urine does not interfere with this test.

If the secretion of urine continues considerably below normal for an extended length of time, other symptoms are pretty sure to follow; the most common of which is headache, either frontal, vertical, or occipital. In chronic uræmia the headache is most commonly occipital, frequently extending down the back of the neck on either side. It may persist for weeks, usually with some intermission during the twenty-four hours. It may be mistaken for rheumatism or neuralgia, if no renal defect is suspected, instances of which have come under my observation. If the symptoms are more threatening, the headache is likely to be frontal. In this location also the pain is more intense and constant as a rule. All grades of severity may be observed as to these headaches; sometimes the pain is excruciating, most commonly so in the supraorbital nerve. As a rule, the more intense and constant the headache, the more imminent the danger of acute uræmia.

In uræmia, disorders of the stomach are scarcely less common than headache, and they are always present sooner or later. They vary in degree from simple loss of appetite, to frequent and uncontrollable vomiting. Most commonly flatulent dyspepsia, and, perhaps, morning nausea are present in the milder cases. More urgent cases are usually attended by vomiting, especially in the morning, though this may occur at any time during the day. The vomited matter is usually of a watery appearance, and may have an ammoniacal odor and alkaline reaction. Diarrhœa is another common expression of chronic uræmia, and is believed to be an effort at vicarious elemination. It is usually associated with disorders of the stomach, such as flatulence and loss of appetite. Distressing itching of the skin is sometimes present. As an illustration of its occasional intensity, Bartels speaks of having seen a patient scratch himself when in a semi-comatose state.

Attacks of dyspnœa are common, and closely resemble asthma in that they often come on at night, are accompanied by sibilant râles, and are followed by more or less expectoration. Insomnia is common—uræmic patients rarely sleep well if the symptoms are at all marked. Transient disturbances of vision are frequent. Uræmic amaurosis consists in a sudden, and more or less complete loss of vision, without any discoverable retinal lesion. It is usually of brief duration, often passing away as suddenly as it appeared. Besides the above, various distortions of vision are common in these cases: thus the patient may see double, objects may appear inverted, or there may be more or less indistinctness of vision.

Deafness, usually of a partial and transient character, is sometimes present. Pains in the muscles, more especially in those of the lower extremities, are common. Twitching of certain sets of muscles, drowsiness, and hebetude of mind are not infrequent. Peculiar attacks of numbness sometimes come on in chronic uræmia. These may be limited to certain nerves, as the facial; or they may include all of one side of the body. They seem to consist of partial sensory paralysis without motor impairment, and are of short duration, usually passing off within an hour. In a case now under treatment, the patient has within the past five months had several attacks, involving the whole left side (hemiplegic). They usually begin with numbness in the left foot, which extends up the side till the left side of the head, including the tongue, becomes affected. The parts remain numb, and more or less insensitive, without motor impairment, for from fifteen minutes to half an hour, when the numbness disappears, leaving no disturbance behind. Lastly, a peculiar form of delirium or intoxication sometimes arises, which may appropriately be termed subacute uræmia. In this condition the patient is restless and uneasy. The eye loses its intelligent expression, and becomes vacant and wild. The patient talks about imaginary events transpiring about him, perhaps calling for friends whom he imagines are present. If spoken to, he usually answers more or less intelligently, and he may even converse rationally for a short time if his attention is closely engaged. A characteristic of this delirium is the more or less complete transient loss of memory. The patient repeatedly asks the same question to which he has just received a reply. He has no recollection as to what transpired a few minutes previously, and when the delirium passes off he is unable to recollect much, if anything, that occurred during the attack. The patient is sensitive to pain during these attacks, and may complain of intense headache, which frequently accompanies it. Great restlessness is present, especially at night: the patient tosses about, talking almost incessantly, and it is very difficult to induce sleep.

The attacks vary in intensity: sometimes the patient shows merely slight mental disorder for an hour or two, which then subsides; or active delirium may be present and continue without intermission. The duration of uræmic intoxication will, of course, depend on the persistence of the cause. It may subside after a few hours' duration, as I have witnessed, with the appearance of free diuresis; it may pass into convulsions, or coma, or both; or it may continue uncomplicated, as in a case that came under my observation, for two weeks, and then subside, the patient remembering nothing of what occurred during the attack.

ACUTE URÆMIA.—Under this head have been classed a remarkable and striking chain of nervo-muscular manifestations, consisting of convulsions, or coma, or both in succession. So far as the nervous system is concerned, the toxic influence seems to be directed (*a*) against the motor centre, the medulla, remarkably increasing reflex irritability of the voluntary muscular system, and causing convulsions, while (*b*) the action on the sensorium and special senses is toward abolition or paralysis of function, resulting in insensibility or coma.

Uræmic convulsions may come on suddenly without warning, the patient being seized while walking in the street, or attending to his business. More commonly, however, they are preceded for some days by some of the symptoms of chronic uræmia already described, more especially headache, disturbances of the stomach, visual disorders, etc. The urine, also, as a rule, is greatly reduced in quantity (oliguresis), and sometimes anuria, or complete suppression, occurs. A single convulsion may occur and subside, passing into coma, or more or less insensibility; or the effect on the sensorium may be transient, the patient regaining consciousness completely; especially is this the case when the convulsion has been mild. More often, however, the convulsions succeed each other at intervals varying from a few minutes to several hours; more or less of the intervals being occupied by partial or complete coma. The convulsions are strikingly epileptic in character, accompanied by distortion of the features, rolling of the eyes, frothing at the mouth; the tongue is liable to be bitten. In the beginning there is strong contraction of the voluntary muscles, commencing usually in the head and upper extremities; on one side at first (usually the left), to which the head is drawn, but gradually extending throughout the whole voluntary muscular system. This is followed by a succession of sudden contractions, or jerks, resembling the muscular contractions produced by an electric battery, which, after continuing for a few seconds, gradually become less severe; muscular relaxation is marked in the intervals. Respiration becomes irregular and embarrassed, and at the height of the paroxysm seems almost suspended. At this point the face often becomes turgid and purple, from interference with the pulmonary circulation; froth mixed with blood escapes between the lips; the globes of the eyes become fixed or rolled upward, and the aspect of the patient is dreadful in the extreme. Gradually, however, if the patient do not succumb to the paroxysm, the contractions become less and less marked, and the intervals lengthen, till at length, in from a few seconds to ten minutes, more or less free diaphoresis ensues, the paroxysm terminates, and a profound stupor succeeds. The temperature is usually exalted, ranging between 101° F. and 105° F. The coma following the convulsions varies in intensity, and is proportionate to their severity and number. The patient is usually profoundly insensible, though often he may be aroused by shaking, or by speaking to him in a loud voice —even to the extent of answering in monosyllables. The respirations become prolonged and stertorous; the noise produced is labial rather than guttural. The blood recedes from the surface, and the patient becomes pale. The pupils may be dilated, but are never contracted. Coma may precede the convulsions, or it may be present without the latter becoming developed at all. If coma occurs alone, it usually develops gradually, occupying several days, perhaps, in reaching the stage of profound insensibility. This is often the method of termination in renal cirrhosis, or lardaceous degeneration of the kidneys. In fact, coma alone is most common in chronic renal disease; while convulsions are most prone to arise in the acute forms of nephritis, such as those caused by cold, scarlatina, and pregnancy.

The patient may succumb during the first or the succeeding convulsive attack; or, surviving these, he may die in the comatose state; but if the convulsions subside, or if the coma be unattended by convulsions, the patient will survive the coma singly much longer than when it is associated with convulsive seizures.

DIAGNOSIS.

In cases in which no renal disease has previously been suspected, uræmic convulsions may be mistaken for epileptic seizures—for the convulsions accompanying both affections are not only very similar, but a like comatose condition follows in each. In uræmia there is more or less pallor of the skin in most cases, due to the renal disease on which it depends. In many cases, also, puffiness of the eyelids, or feet, is observed; and, as a rule, to which there are comparatively few exceptions, the urine is albuminous. In uræmia, the attacks are usually preceded by more or less drowsiness, headache, visual disturbances, nausea, vomiting, etc. In uræmic coma the patient is capable of being aroused, and made to obey commands, or to answer questions; the stertorous sounds of the respiration are made in the mouth rather than in the throat; and there is pronounced irritability of the muscles on percussion. Uræmic attacks are most liable to occur during or after muscular exercise, when the patient is about or unusually active; while epilepsy often occurs at night, or during muscular quietude. The history usually points to previous attacks in cases of epilepsy; while in uræmia it will be almost impossible to overlook the renal condition upon which it depends. Lastly, a succession of convulsions, with brief intervals, are much more frequent in uræmia than in epilepsy.

The coma of uræmia may be mistaken for apoplexy, and this deserves careful consideration from the fact that both are not infrequently the outgrowth of the same primary cause; thus presenting very similar histories. In typical hemorrhagic apoplexy there are no true convulsive movements, the patient falls down in a state of insensibility, which usually deepens till it is impossible to arouse him. The respirations become stertorous, with deep guttural tones; the skin is normal in color and temperature. The pulse is slow, 60 or less in the minute, full in volume, and rather labored. Paralysis is the rule in apoplexy; it is usually complete, and of the hemiplegic variety. The tongue is only exceptionally bitten, and it is never deeply lacerated, as in epilepsy or uræmia; frothing at the mouth is seldom observed. In apoplexy the pupils are usually unequal.

Cases may occur in which a person is found in a state of insensibility, the preceding symptoms having been unobserved, and a physician may be called upon to decide between uræmic coma and the stupor of alcoh8lism. Or uræmic coma may occur in the case of an intemperate individual, and be preceded by mental intoxication, and thus lead the friends to believe the condition that of alcoholism. The odor of alcohol in the breath may properly arouse suspicion, to which further inquiry may be directed, but this must not be relied upon as positive evidence of drunkenness, for the subjects of Bright's disease and uræmia are not infrequently addicted to intemperate habits. The pallor of skin and puffiness of the face, already spoken of as common accompaniments of uræmia, if observed, will lead the inquiry in the correct direction. Reflex muscular irritation and sensibility to pain are more pronounced in uræmia than in alcoholism. If the patient be partially aroused and made to answer brief questions, the articulation is found more distinct in the case of uræmia; but if, after careful examination, there remain any doubt, it will be best to draw off some urine with a catheter, and test it—the absence of albumin and other symptoms of nephritis, is pretty conclusive evidence that the case is not one of uræmia.

PROGNOSIS.

In acute uræmia, whether in the form of convulsions or of coma, or, as is most common, both associated together, the prognosis must be considered as dangerous to life— but the greater danger arises from convulsions. The mortality in those attacked with puerperal convulsions is placed by Braun at thirty per cent. This estimate is probably higher than the actual mortality from convulsions in other forms of acute Bright's disease, while it is probably below the actual figures in convulsions of chronic Bright's disease, as the conditions causing the latter are less remediable by medication. From this it will appear that the prognosis, in a given case of uræmic convulsions, is influenced by the nature of the renal lesion upon which the convulsions depend, especially as regards ultimate recovery. As to the attack itself, we shall be able to form some idea as to the probability of the patient's survival, from the severity, duration, and frequency of repetition of the paroxysm; the condition of the patient during the intervals, and whether secondary complications are present or not, etc. Thus frequent, violent, and prolonged paroxysms, resulting in profound unconsciousness which continues more or less completely during the intervals; the occurrence of serous exudation into the lungs with consequent dyspnœa, or into the brain or its ventricles; marked exaltation of the pulse and temperature, point strongly toward a fatal termination of the case. On the other hand, if the violence, length, and frequency of the convulsions diminish to a marked extent; if no secondary disease of the brain or lungs has been caused; if the pulse become normal and quiet, the temperature lowered, and free diuresis established, the prognosis assumes a favorable phase.

In uræmic coma unassociated with convulsions, the prognosis is more favorable as regards the immediate result, than in coma with convulsions—as more time is afforded to secure the effects of remedial measures. The ultimate prognosis, however, is more gloomy in the former than in the latter case, as simple uræmic coma is usually a manifestation of incurable chronic lesions of the kidneys. Chronic uræmia presents a more encouraging outlook—being largely influenced by well-directed medication. The ultimate result, however, depends on the character of the renal lesion.

TREATMENT.

In chronic uræmia, the symptoms are less urgent than in the acute form, and more time is afforded in which to restore the renal function—and thus open up the natural channels for escape of the retained waste elements, which are slowly damaging the organism.

Upon the first indications of uræmia—such as headache, nausea, and oliguresis—the patient must be strictly confined to his room, or, better still, his bed, as muscular exercise aggravates the symptoms, and may even be the means of provoking an acute attack.

The diet must be strictly regulated, so as to contain a minimum of nitrogenous elements—as these go to make up the waste that passes out by the kidneys—thus diminishing the work thrown upon those organs.

The matter of atmospheric surroundings of the patient deserves more consideration than it has hitherto received—both because elimination by the lungs is a factor of no small importance, and because pure air very materially increases the oxidizing powers of the blood so essential to healthy metamorphosis of tissue. The patient should be assigned a large, well-ventilated room, and the temperature should be as low as is compatible with comfort.

As to medication, the first measures to be employed should be such as are calculated to promote diuresis. As the appropriate selection of these will depend largely on the nature of the pathological conditions which interfere with the excretion of the urine, they will be more minutely dealt with in the sections on special forms of renal disease. It may, however, be generally stated here that in the selection of agents to be employed—where no contraindications exist—a union of diuretics is most useful, combining those that favor filtration, with those that stimulate the renal cells in removing solids—thus securing more thorough depuration of the blood, as recently shown by Dr. Fox (see *Brit. Med. Journ.*, Aug. 22, 1885). The combinations of digitalis with neutral salts of potassium, sodium, or lithium, best answer the purposes in acute cases, as they possess the fewest contraindications. In non-inflammatory states of the kidneys, we may employ agents that both stimulate the renal cells and act as vaso-dilators on the renal vessels; such as juniper, squill, scoparium, etc. Special care must be taken, however, that the latter are not employed in acute congestive conditions of the renal organs.

What means have we at command that are capable of lessening the force of the toxic influence exerted upon the organism in these cases? Aside from those that act as repressants upon the nervous system—preventing or lessening the intensity of the convulsions, I am aware of but two measures that directly modify the toxicity of the system. The first is subcarbonate of iron given in large doses—twenty to thirty grains—and repeated every three hours. In chronic uræmia, where we have time to secure the general effects of medicines, I have found decided benefit from this remedy. It usually rapidly relieves the headache, nausea, and other symptoms—possibly acting as an oxidizing agent on the effete materials with which the blood is surcharged. The subcarbonate of iron is well adapted for administration, as in the majority of these cases anæmia is marked.

The second method of modifying the uræmic influence is by simply diluting the blood by means of copious draughts of fluid. It is apparent that the more dilute are rendered the effete products exerting the toxic influence on the organism, the less intense must be their effects. The ingestion of fluids exerts still another influence in these cases: by increasing the volume and fluidity of the blood, filtration and diuresis are promoted. The eliminative treatment of uræmia will be found the most effective of our resources, and it has proven so beneficial in the past, that there remains little room for argument on the question. This method aims at getting rid of the accumulation of waste products and toxic matters through the excretory channels.

The skin deserves special consideration in this connection, as it is capable of being stimulated into very free eliminative action, under the influence of certain agents. The most efficient of these are hot baths and jaborandi, or the active principle of the latter—pilocarpine. The most thorough action obtainable, is by means of *dry* hot air. Hot vapor or water baths, or the wet pack, are serviceable and often convenient methods of exciting the action of the skin, but they are not so efficient as dry hot-air baths, for the reason that evaporation from the skin is vastly more rapid in dry than in moist air. Of scarcely less importance is the fact that the bodily temperature is by no means so increased by dry, as moist air, for obvious reasons. The action of these diaphoretic measures is much increased by the use of hot drinks. The hot-air baths may conveniently be employed by seating the patient in a chair with a perforated seat, under which an alcohol lamp is placed, the flame being covered by a piece of sheet-iron to protect the patient and diffuse the heat. Around the patient and chair are wrapped several thicknesses of blankets, to retain the heat. Profuse diaphoresis may thus be induced and maintained as long as desired. Still more comfortable to the patient, especially if weakness is marked, is to administer the hot-air bath while the patient is lying in bed. To accomplish this, I am in the habit of using a hot-air apparatus that I have had constructed for the purpose. This consists of a sheet-iron box, twelve inches square, and nine inches in height, perforated at the bottom, and fitted at the top with several joints of pipe a foot and a half in length, and two inches in diameter, one link of which contains an elbow. An alcohol lamp is placed in the box, over the flame of which a diaphragm is arranged, which acts as a radiator. By means of the pipe, the heat is conducted into the bed beside the patient; the box being placed at sufficient distance from the bed to protect the patient from too intense heat. The apparatus is simple and portable, and may be constructed by any tinsmith in an hour or two. By means of such an apparatus, I am in the habit of producing and maintaining free diaphoresis for an hour or more, while the thermometer under the tongue never marks a rise exceeding one degree, and often not over half a degree. When the hot water or hot vapor bath is used, the temperature invariably rises from two to four degrees, which much increases the risks of acute uræmia following, if the symptoms are threatening. After the hot-air apparatus is removed, if the patient be given a warm drink and kept covered, diaphoresis may be kept up for an hour or two, if desired.

Jaborandi is an exceedingly efficient diaphoretic, and it possesses the advantage over all hot baths, that it does not increase the bodily temperature, and thus precipitate acute uræmia if the latter be threatening. In fact, when sweating is produced by it, a distinct fall of temperature ($0.5°$ to $2°$ F.) ensues, and this decline of body heat is maintained on an average about four and a half hours (Bartholow). Jaborandi may be administered in the form of fluid extract, in doses of a half to two drachms; or its active principle, hydrochlorate of pilacarpine, may be given in doses of one-twelfth to half a grain. The most rapid action is secured by subcutaneous injections of the pilocarpine hydrochlorate, in doses of one-tenth to one-eighth of a grain. Jaborandi possesses two or three effects undesirable in certain conditions, which it is necessary to bear in mind. It frequently depresses the heart's action, and may produce marked prostration; hence its use should, if possible, be avoided in patients with dilated and weak hearts—such as are liable to be encountered in scarlatina, or the late stages of renal cirrhosis. Jaborandi also occasionally induces respiratory embarrassment of a congestive and œdematous nature—and this fact deserves the more attention in consequence of the fact that bronchial and pulmonary complications are peculiarly liable to arise during renal disease.

I have witnessed the production of œdema of the lungs in several of these cases by the use of this agent—in one instance with fatal result. I, therefore, use jaborandi with extreme caution in cases of uræmia; and never, if the respirations are abnormal. Lastly, jaborandi is said to stimulate the gravid uterus, and cases are recorded of its having induced premature labor—which effect should not be lost sight of in treating eclampsia.

We possess in purgative medicines very valuable agents to relieve uræmic conditions. The intestinal tract is capable of being readily stimulated into great eliminative activity. If necessary, and the general condition of the patient permit, an enormous drain of fluid may be brought about through this channel—carrying with it in solution much of the effete matters and extractives giving rise to the uræmic disturbance. The choice of the special cathartic should depend on the urgency of the symptoms; but the best results are obtained in all cases from the action of hydragogues. Elaterium has heretofore borne the highest reputation in these cases. Owing to the uncertainty regarding the strength of this agent, its active principle, elaterin, should be employed instead—dose, one-twentieth to one-eighth grain; this may be repeated every four hours until free watery stools are produced. In many cases elaterin is admirably adapted for the purpose under consideration. Not infrequently, however, this agent produces vomiting, or sharp irritation of the intestinal tract. Unfortunately also, in a large number of renal diseases causing uræmia, the bowels are already more or less irritated. Under such circumstances it becomes desirable to use intestinal eliminants that will not aggravate existing irritation. In such cases I have found concentrated saline solutions, as recommended by Dr. Matthew Hay, to meet the requirements most satisfactorily. The best results are obtained from the administration of an ounce of sulphate of magnesia, or Rochelle salts, in as small a quantity of water as will hold it in solution, say an ounce and a half. The patient should take no fluids into the stomach for twelve hours before, or six hours after, taking the saline; hence it is best administered early in the morning. Given as directed, the salines will bring away enormous quantities of fluid from the intestines, attended by but little intestinal irritation and systemic disturbance.

Muscarine has been advised as an eliminant in these cases; besides acting on the bowels, it also stimulates the secretion of the pancreas, liver, salivary glands, and skin—the latter only less actively than jaborandi. It is sedative in large doses, and must be used with caution in case of weak heart. Its dose is from one-eighth to two grains, and it possesses the advantage of acting quite as efficiently when given subcutaneously as when given by the stomach. A one per cent. solution of the extract may be given, in from five to ten minim doses. In less urgent cases, the compound jalap powder is much employed; dose, ten grains to ʒj.

There is one danger it is well to guard against in the eliminative treatment of uræmia under certain conditions, and that is dropsy. Considerable quantities of the toxic matters which are believed to be the cause of uræmia often become stored up in the dropsical fluids in the great cavities and cellular tissues of the body; and if the dropsy be rapidly reduced through purgation or other eliminative means, the matters thus stored up are thrown suddenly into the circulation, and are liable thus to precipitate convulsions, or coma. The danger is not an imaginary one, for it has been verified by several observers, and in one instance I witnessed the production of uræmic intoxication by this means. I know of but one way to lessen this danger—and that is to give copious draughts of fluids while the eliminative measures are being carried out. The blood is thus repleted from the fluids ingested, and a true depurative action is obtained.

Should uræmia assume an acute type, and convulsions appear, additional measures must be resorted to in order to preserve the life of the patient. As to the management of the convulsive seizures themselves : if the patient is seen in their beginning, it is well to insert some soft substance, such as a cork, between the teeth to protect the tongue from laceration ; next, to secure free access of pure air, as the respirations frequently become greatly embarrassed (almost arrested) at the height of the paroxysm, inhalations of chloroform, or ether, may next be used, as they modify the violence of the muscular contractions. The former is the more efficient, and must be used with caution. As soon as the convulsions pass off, a full dose of chloral hydrate—thirty to forty grains—should be given, preferably in peppermint-water, or in a little port wine. If this be successful in postponing the attacks, the impression should be kept up by the administration of smaller doses—fifteen to twenty grains—repeated every six to eight hours.

If the convulsions are violent and occur in rapid succession, and the general condition of the patient is good, not having been previously enfeebled by chronic disease, we should not hesitate to resort to venesection. Accumulated experience no longer permits us to doubt that many lives have been saved by this measure. It should be resorted to boldly in order to secure its best effects—twelve to twenty ounces should be taken at once in the case of adults. Unfortunately, few of the cases coming under observation are in sufficiently good condition to render the employment of this measure permissible—by far the greater number being more or less reduced from long-existing renal disease. Morphine has been highly recommended for the relief of uræmic convulsions. It is said that its use in these cases was originally practised by Scanzoni. A strong prejudice has existed against the use of opiates in these cases, owing to the fact that even small doses have been known to cause death when administered to patients with diseased kidneys. But we are indebted to Dr Loomis, of New York, for having demonstrated that morphine can be administered subcutaneously in cases of uræmic convulsions, not only without endangering life, but often with the happiest results. He writes (*Diseases of the Respiratory Organs, Heart, and Kidneys*, third edition, 1882): "From the histories of quite a large number of puerperal and non-puerperal cases of acute uræmia in which morphine was successfully used, I have reached the following conclusions:

"*First.* That morphine can be administered hypodermically to some, if not to all, patients with acute uræmia, without endangering life.

"*Second.* That the almost uniform effect of morphine so administered is first to arrest muscular spasms by counteracting the effect of the uræmic poison on the nerve-centres; second, to establish profuse diaphoresis; third, to facilitate the action of cathartics and diuretics, especially the diuretic action of digitalis."

The range of utility and safety of morphine in uræmia is strictly confined to the convulsive seizures; its employment is improper in the state of pure coma, or chronic uræmia unassociated with convulsions.

In acute uræmia the functions of elimination are usually held in check, and Dr. Loomis administers morphine under such circumstances with the view to control the convulsions until these functions shall be restored—to "hold the patient until the normal eliminating process shall be reëstablished."

Others have since corroborated the statement of Dr. Loomis as to the beneficial action of morphine when administered subcutaneously in cases of uræmic convulsions. It is, therefore, our duty to resort to the use of this measure in these cases, regardless of any prejudice we may have against the administration of opiates to patients whose kidneys are diseased; and more especially is this course advisable if the convulsions do not yield to other remedies.

It is better to begin with small doses of morphine (onesixth to one-fourth of a grain), and repeat, if necessary, than to administer the larger doses at once.

In the intervals between the convulsions, the measures already advised for the relief of chronic uræmia should be actively pushed. Every effort should be made to reëstablish the functions of elimination. Digitalis is especially serviceable throughout. Active cathartics should be given without hesitation, as these are usually very efficient aids, not only on account of their eliminative action, but also because they lessen the congestion of the nerve centres, and so diminish the risks of permanent damage to the brain.

While such measures as have been considered are being carried out, the exact pathological condition of the kidneys which has caused the uræmia should be ascertained, if possible. Appropriate measures must be instituted for the relief of this so far as is possible; for on our success in this direction will depend the permanency of relief to the uræmic symptoms. It has already been shown that uræmia is caused by failure to excrete the urine or its derivatives from the system; and consequently we can expect no permanent improvement until the function of the kidneys is restored. The means of accomplishing this under the various pathological conditions of the organs will be next considered in detail.

CHAPTER III.

ACUTE NEPHRITIS.

ACUTE inflammation of the kidneys has been described under various names, such as acute tubal nephritis; croupous nephritis, acute desquamative nephritis, acute catarrhal nephritis, acute parenchymatous nephritis, acute Bright's disease, etc. As most of these terms are restrictive in their signification, implying only part of the morbid process and pathological changes progressing in the organs, they are likely to mislead the student; therefore, a name which implies only the type of the disease has been selected as the most appropriate.

ETIOLOGY.

Acute inflammation of the kidneys is most frequently observed in young adults; though it is not uncommon, also, in middle age. Like most other renal inflammations—the puerperal form, of course, excepted—it occurs about three times more frequently in the male than in the female sex.

It is much more common among the middle and lower classes, because their mode of life brings them in frequent contact with the most prominent cause of the disease.

Exposure to Cold.—Slight and transient exposures to cold are seldom sufficient to induce an attack. It more often results from chilling of the skin through protracted exposure to a moist, cold atmosphere. Exposure to cold is more likely to excite the disease in those who are exhausted, or debilitated, or in those whose skin is in an active state of perspiration at the time of exposure. This is not difficult to comprehend, since we know that the effects of exposure to cold are always most profound under these circumstances; the vaso-motor system is depressed, and less responsive to external influences, such as cold; and the cutaneous vessels, through their temporary inability to contract, become distended with blood, and the latter receives the full force of the influence of cold atmospheric currents.

The disease, therefore, arises most commonly from exposure when the body has been overheated by unusual exercise. In the laboring classes we see it oftenest in those whose work is irregular in character; those, for instance, who are actively employed for a time, and then stand about exposed to cold winds while in a state of perspiration. In the higher classes the affection is usually met with in those who, after a night of dancing or dissipation, lie down in chilly apartments and sleep without sufficient covering.

Dr. Ralfe (*Diseases of the Kidneys and Urinary Derangements*, 1885) has called attention to the singular circumstance that exposure to cold rarely gives rise to acute nephritis in children. His experience as physician to the London Hospital must have brought him in contact with numerous cases of exposure in children in the East End of London, yet he says, "I have never yet succeeded in obtaining a history of exposure to cold and wet in a case of acute nephritis occurring in childhood."

Febrile Diseases. — Certain diseases often give rise to acute nephritis, such as diphtheria, smallpox, measles, erysipelas, typhoid fever, and pneumonia. The nephritis excited by these diseases is usually mild in character, of brief duration, and is chiefly limited to the pyrexial stage. Exceptional instances occur in which the disease attains a more active stage from these causes; being accompanied by dropsy and bloody urine, etc. As a rule, however, the symptoms are limited to more or less albuminuria, and that chiefly in the pyrexial stage, as already stated.

Toxic Agents.—Certain substances introduced into the blood, which are eliminated by the kidneys, are capable of giving rise to acute nephritis, through their irritating effects on the renal organs when taken in large doses, such as cantharides, turpentine, copaiba, phosphorus, and alcohol. Some of these act as violent irritants, even in comparatively small doses under certain conditions, notably cantharides. Phosphorus induces fatty degeneration in the renal epithelium. The etiological relations of alcohol to nephritis has long been a debatable question. Since the experiments of Penzoldt, however, there can be no further doubts of its capability of producing albuminuria and nephritis A number of other agents act as irritants to the kidneys when introduced into the blood, as arsenic, mercury, antimony, nitrate of potassium, salicylic acid, etc. It is rarely, however, that the bounds of simple congestion are overstepped as the result of the action of these, though exceptionally cases of nephritis have been traced to their influence.

Cutaneous Lesions.—Extensive destruction of the cutaneous structures through disease, but more especially when resulting from burns, sometimes excites acute nephritis. Several explanations of this have been advanced, none of which appears quite satisfactory. Bartels says destruction of the skin, if extensive, "causes a general depression of the temperature of the body in consequence of the great loss of heat; it acts in the same way as continued abstraction by cooling of the uninjured skin." Ralfe thinks that nephritis is caused in these cases by interference with the cutaneous function, resulting in "the non-elimination and consequent retention in the blood of deleterious excretory products."

Of the various causes of the disease just considered, exposure to cold is the most fertile one; and probably a majority of cases met with in practice owe their origin to this source.

MORBID ANATOMY.

If the kidneys be examined after death from typical acute nephritis, they will be found more or less enlarged, sometimes very much so; and their shape more rounded than normal. The surface of the organ is smooth, pale in color, and mottled by irregular yellow spots. The organs are less firm than normal, feeling soft and flabby to the fingers.

The capsule is non-adherent; in fact, it strips off more easily than normal, and it is undermined by œdematous tissue. No cysts are observed on the surface, except old ones due to previous chronic disease.

On section of the organ it is seen that the cortex is much thickened, being to the medulla relatively as one to two, or even equal to the latter, especially at parts which dip in between the pyramids. Although the vessels contain much blood, when the latter is washed away the cortex appears paler than normal, and of the same mottled appearance as the surface of the organs. Small red points may be seen scattered over the cut surface of the cortex, more particularly near the surface, which are enlarged and deeply congested Malpighian bodies. In cases in which the kidney is greatly enlarged, the deep prolongations of

90   ACUTE NEPHRITIS.

the cortex become so much swollen that they constrict the pyramids between them; producing the appearance which Rayer described as like that of a "sheaf of wheat." The medulla and mucous membrane of the renal pelvis are considerably congested.

MICROSCOPIC.—On microscopic examination very pronounced changes are observed, which are chiefly confined

FIG. 2.

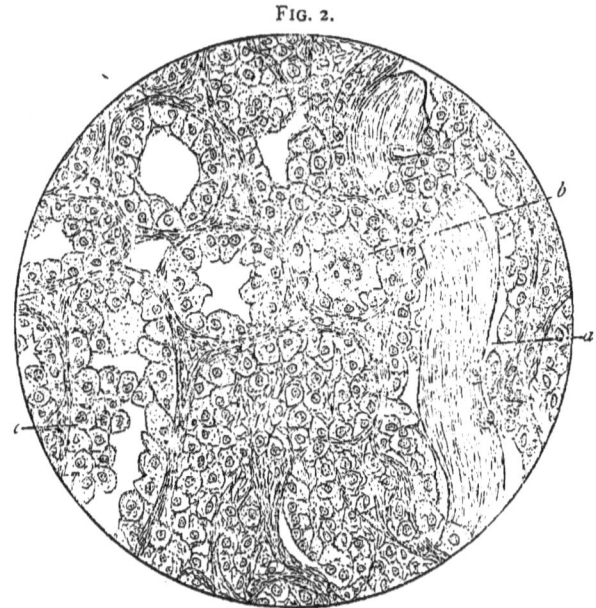

Acute nephritis. Transverse section through cortex. *a*. Hyaline cast within convoluted tube. *b*. Cross-section of tubuli containing granular cyst. *c*. Degenerated and exfoliated epithelium seen in longitudinal section of tubule.

to the cortex. The convoluted tubes are much swollen; their epithelial lining is in various stages of degeneration. In some cases the epithelium is merely in a typical state of cloudy swelling; in others, necrosis and breaking down of

MORBID ANATOMY. 91

the cells into granular elements are plainly in progress; and in many tubes the epithelium is infiltrated with fatty matter. Some tubes are seen with the epithelium fallen out; and in all places where remaining the epithelium is swollen, and undergoing degenerative changes of a granular or fatty character. The convoluted tubes are seen to

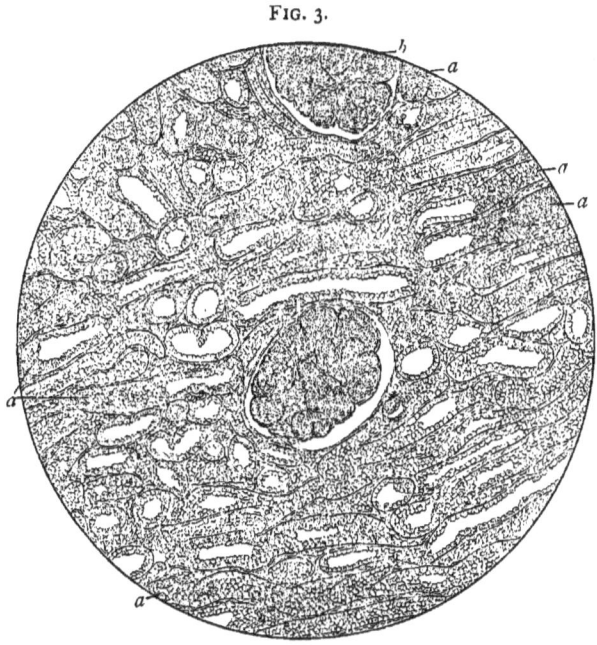

FIG. 3.

Acute nephritis. Section of cortex. *a.* Fatty degeneration of tubular epithelium (drawn from an osmic acid stained specimen). *b.* Malpighian tuft.

be filled with the elements of broken-down epithelium, blood-corpuscles, minute oil-globules, and colloid material, as far down as Henle's loops. This débris engages at Henle's loops, and chokes up the tubes owing to the narrowing of the latter at this point; hence it accumulates in the cortex. If the débris gets into the medulla, it works out of the straight tubes.

The interlobular vessels are much distended with blood, and between the convoluted tubes and around the Malpighian bodies patches of blood-corpuscles (hemorrhages) may be seen. Sometimes close to the capsule of Bowman may be observed an increase of minute points, which are nuclei of young connective tissue cells, showing that interstitial changes have been initiated.

The glomeruli are seen to be enlarged, their tufts swollen, and filling the capsules of Bowman. The epithelium lining the capsule of Bowman, as well as that covering the tuft, is swollen, granular, and desquamating; but little fatty degeneration is seen, like that which is so extensively found in the epithelium lining the convoluted tube. Within the capsule of Bowman are frequently seen numerous blood-corpuscles (hemorrhages) which result from rupture of vessels of the tuft.

The straight tubes in the medulla are little, if any, affected, their epithelium remaining unchanged.

Such are the typical changes found in the kidneys resulting from acute nephritis. They correspond with the clinical symptoms associated with pronounced oliguria (or suppression); the urine being more or less bloody, and rapidly developing dropsy.

The process is essentially acute inflammation involving the whole glandular structure or cortex. It is true that comparatively slight changes are noted in the connective tissue of the organ, because all marked changes in such structures require more time for their development than is occupied usually by acute nephritis, yet even here the increased nuclei of young connective tissue cells are sufficiently observable, especially around the capsule of Bowman, and nearest the surface of the organs, to show

SYMPTOMS.

that changes in the connective tissue have been inaugurated; and the longer the disease continues the more marked these changes become.

SYMPTOMS.

The symptoms of acute nephritis usually begin with a more or less pronounced chill, followed by pyrexia for a few days, during which the temperature may range from 100° to 104° F. The pulse is usually increased in frequency and volume; the tongue is more or less coated, and the bowels constipated. In some cases these symptoms are greatly modified, although rarely altogether absent.

The Urine.— This is always diminished in quantity, sometimes completely suppressed. More often, however, the quantity ranges from four to sixteen ounces in twenty-four hours, during the height of the disease. The color is reddish-brown, and it is more or less turbid or "smoky" from admixture with blood and cellular elements. The specific gravity is increased, ranging between 1.030 and 1.040 when the symptoms are acute and the quantity small. Later, as the quantity increases, the specific gravity falls. The reaction is usually acid.

There is a disposition on the part of the patient to urinate frequently, sometimes passing but a few drops at a time—vesical tenesmus.

Uneasiness in the loins may be felt, sometimes amounting to pain, although the latter is more frequent along the course of the ureters, and in the testes, or in the glans penis immediately preceding or following the act of micturition.

The urine always contains albumin, usually in large quantities, ranging from 0.5 to 3 per cent. by actual weight; or when coagulated in the test-tube it occupies from one-third to three-fourths of the latter after settling.

Urea is decreased in quantity from twenty-five to fifty per cent., and in some cases it may be absent. The remaining solids are all more or less reduced, but the chlorides the most so.

The sediment, which is more abundant than in any other renal disease, is made up of casts, blood corpuscles, epithelium, and granular urates. In some cases casts and epithelium predominate, and in others the sediment consists principally of blood corpuscles. More or less blood corpuscles and epithelium are nearly always present in the urine throughout the disease. In the early stage the epithelial and blood casts are the most numerous; later, granular and fatty casts become common, indicating a transition into a more chronic condition.

Dropsy.—This is one of the most constant symptoms of the disease, and it is often the first to attract the attention of the patient. First, it may be observed that the eyelids or cheeks are puffy in the morning on rising; or the feet may be swollen toward evening, especially if the patient has been up and about. Thus, beginning with less marked symptoms during the first two or three days, the dropsy usually increases quite rapidly, until not only the whole subcutaneous tissue throughout the body becomes more or less infiltrated, but also the large sacs—peritoneal, pleural, and even pericardial—become more or less involved, if the disease continue unabated. More frequently, however, the subcutaneous tissue and the abdominal cavity are the principal seats of the serous effusion. The limbs sometimes become enormously swollen; as do sometimes the eyelids, face, scrotum, and penis; the penis may become bent and distorted in a most grotesque manner. This disfiguration of the genital organs often greatly disquiets the mind of the patient. But more serious consequences are likely to ensue from the accumlation of fluid within the thoracic cavity. The pericardial or pleural effusion may embarrass the action of the heart or lungs to such an extent as to threaten life. The glottis occasionally becomes œdematous ; threatening life from asphyxia, though this complication is fortunately rare.

Digestive System.—The appetite is lost, especially during the pyrexial stage; and in most cases this continues till late in the disease. Nausea and vomiting are common. In the beginning they are probably purely reflex; later they are of uræmic origin, and may point to symptoms of impending gravity. The bowels are usually constipated in the pyrexial stage; later, diarrhœa may set in, which is probably an eliminative effort of nature to get rid of accumulating noxious products.

The Circulatory System.—Hypertrophy of the heart is uncommon in this disease, as cardiac changes of a hypertrophic nature require more time for their development than is usually afforded by the duration of acute nephritis. Exceptionally, acute enlargment of the left ventricle may become developed in a mild degree, but this usually subsides quickly on convalescence if the disease terminates favorably. There is no doubt that the work of the heart is abnormally increased in this disease, as evidences of increased arterial pressure are often present, especially in the early course of the disease; as a rule, however, these subside as the pyrexial stage passes off, or sometimes before.

Anæmia is an early and marked characteristic in these cases. Indeed, nothing so quickly arrests the attention of the experienced observer, and leads him to suspect the presence of this disease as the blanched and pallid features of the patient. This early anæmia has been explained as the result of the loss of albumin through the kidneys; this is no doubt considerable in most cases. I must, however, agree with Bartels (op. cit.) that the renal hemorrhage in these cases is the active agent in bringing about anæmia. From the fact that the acids of the urine darken the blood before it escapes from the bladder, I think the extent of the renal hemorrhage is not always appreciated, as it is not always apparent. If, however, the sediment be examined under a microscope in the active stage of the disease, it will be found composed largely of corpuscles, and, consequently, the hemorrhage from the kidneys in some of these cases must be very considerable.

The blood itself suffers marked change in acute nephritis; its corpuscles are reduced in number, its albumin is diminished, and its aqueous elements are increased—hydræmia. In addition to this, urea and uric acid accumulate to a greater or less extent within it, together with the various extractives, or urinary derivatives.

Hemorrhages into the retina, or in other locations, as epistaxis, are rare in acute nephritis, both because the coats of the small vessels remain unchanged, and the heart rarely attains a marked degree of hypertrophy. Lastly, pericarditis is not infrequent in acute nephritis, and this is often attended by purulent effusion into the pericardial sac, nearly always resulting fatally.

The Nervous System.—The most marked effect of this disease upon the nervous system is the production of uræmic convulsions. Coma is less frequent unless associated with convulsions, in which case it may occupy more or less of the intervals between the convulsive paroxysms. Headache is rarely absent during the pyrexial stage, or when the urine is scant and bloody. The pain is more or less intense according to the activity of the symptoms;

sometimes it is excruciating. Uræmic amaurosis occasionally occurs, resulting in more or less complete, though transient, blindness; permanent impairment of vision is rare.

The Respiratory System.—Bronchitis and pneumonia sometimes occur, the latter being a very fatal complication. Pleurisy is a very frequent complication, and is usually fatal. This is due to the fact that it is commonly accompanied by purulent effusion into the pleural cavity, which almost necessarily places the case beyond hope.

The remaining complications of the disease are peritonitis· and gangrenous affections of the skin. The latter are much less common than in chronic renal disease.

COURSE AND DURATION.—The first week or ten days may be reckoned as the pyrexial stage of the disease, in which the symptoms are subject to great variation. The temperature during this period may range between 100° F. and 104° F., or it may rise but little above normal. Dropsy sets in early and is in inverse ratio to the quantity of urine passed. The more extreme grades of dropsy are usually reached after ten days or a fortnight's duration of the disease. The urine varies greatly both in quantity and quality. If the attack be acute, the quantity of urine is diminished to a few ounces and this is loaded with albumin, blood corpuscles, and casts. Later if the case assumes a favorable phase, the quantity of the urine becomes increased; the blood corpuscles are diminished or even absent; the specific gravity is lowered, and the amount of albumin is decreased.

Should the more acute symptoms continue unabated, especially great diminution in the quantity of urine, the latter containing much blood, some of the secondary complications are liable to arise during the second or third week, or even earlier; or acute uræmia may supervene. Any of these complications may quickly prove fatal.

Sometimes the disease develops more gradually, the pyrexial symptoms being absent or unnoticed in the beginning. Later, the more acute symptoms may become developed, as anuria, dropsy, secondary inflammations, uræmia, etc.

The disease may pursue still another course. After the acute symptoms have to some extent subsided, a relapse may occur, through exposure to cold, undue exercise, or improper diet. It should be borne in mind that relapses are likely to occur at any time during the course of disease from carelessness on the part of the patient; and they bring with them all the dangers common to an acute attack.

The disease may assume a favorable phase at any period in its course, and this may continue until complete recovery results. In such cases the urine becomes increased in quantity—as a rule, considerably above normal; the specific gravity becomes lowered; the albumin decreases in quantity; hæmaturia subsides; the dropsy is reduced; casts become less numerous; and thus convalescence which may terminate in recovery becomes established.

In less favorable cases the symptoms improve up to a certain point, the quantity of the urine becomes more or less increased; the blood disappears from the urine; albumin diminishes in quantity, but does not completely disappear; casts are still observed though in less number, and now more granular and fatty ones are present; dropsy may or may not subside; and thus the disease may pass into the chronic form and continue indefinitely.

In cases that pursue a favorable course, improvement usually begins during the latter part of the first week;

and if the improvement is uninterrupted the urine will rapidly undergo changes the tendency of which is toward the normal. In such cases both albumin and casts may completely disappear in from two to six weeks; but casts may usually be found for a longer or shorter time after the disappearance of albumin.

In other cases the acute symptoms are not developed till later, or relapses may occur; or the acute symptoms may only partly yield, and the disease may be prolonged to eight or ten weeks.

As a rule, however, if the disease extend beyond six to eight weeks the symptoms becoming less urgent, hæmaturia subsiding, the quantity of urine approaching the normal, but albumin and casts remaining, the disease must be considered as having passed into a chronic state.

DIAGNOSIS.

CONSTRUCTIVE.—The recognition of acute nephritis presents fewer difficulties, perhaps, than that of any other renal disease. The symptoms are so marked, even at an early date, that the patient's attention becomes directed to the morbid phenomena, and the case comes under observation comparatively early in the majority of instances. Dropsy is rarely absent when the disease is established. With this we have diminution in the quantity of urine; and the urine contains blood-disks, sometimes in large numbers, epithelial casts, degenerated epithelium, and a large amount of albumin. The urine is dark in color, turbid, and "smoky" in appearance, of high specific gravity—over 1.020, and deposits much sediment.

The general appearance of the patient is anæmic, with puffy features or extremities, or both. A history of recent exposure to cold often completes the circle of characteristics, which are usually sufficiently distinct to render the diagnosis clear.

DIFFERENTIAL.—Acute nephritis may be mistaken for cyanotic induration of the kidneys, as the latter disease presents some points of similarity to the former, such as diminution in the quantity of the urine, which is of high specific gravity and of dark color, and the presence of more or less dropsy. The distinguishing points to be kept in view are, the small amount of albumin in the urine; the sparse casts, which are usually hyaline; dropsy confined mostly to the lower extremities; the presence of more or less dyspnœa; marked cyanosis of the skin; evidences of valvular lesions in the heart, characteristic of cyanotic induration of the kidneys. These symptoms are essentially different from those present in acute nephritis. Acute nephritis sometimes complicates chronic renal disease, and it is then often difficult to distinguish by the symptoms alone the nature of the original lesion. In such cases we must be governed largely by the history. If renal disease has been in progress for a considerable length of time, evidences of its presence may usually be elicited upon careful inquiry.

Albuminuria sometimes arises in the course of febrile diseases, and the albumin may become so considerable in quantity that acute nephritis may be suspected. In such cases there is, undoubtedly, more or less congestion of the kidneys; but the symptoms are always much milder than in acute nephritis. The quantity of urine is rarely diminished to the extent seen in acute nephritis. The casts and sediment are less, and blood in the urine is rare. In albuminuria of the febrile state dropsy is absent, anæmia is not marked, and the albuminuria usually disappears with the subsidence of the pyrexial symptoms.

PROGNOSIS.

The majority of cases of acute nephritis terminate in complete recovery. Frerichs estimated the proportion of recoveries at two-thirds of those attacked. It is probable that a larger proportion than this would end in recovery if subjected to proper treatment in the beginning. Dr. Roberts (*Urinary and Renal Diseases*, 1885) has observed that the prognosis is more favorable in the aged than in the young. In his experience, the disease is mild in persons over sixty, and always ends in recovery.

During the early course of the disease the symptoms which may be looked upon as unfavorable are, first and most prominently, more or less complete anuria. Complete suppression is comparatively rare, but the degree of diminution in the quantity of urine is the best index of threatening danger. If the urine continue small in quantity, and contain blood in considerable proportion, it is an unfavorable indication. Secondary inflammations within the chest, more especially those associated with purulent effusion, are almost necessarily fatal. Less fatal, yet still a most grave complication arising in the course of the disease, is the appearance of convulsions.

If, on the contrary, the quantity of urine reach or exceed the normal, the hæmaturia subside, the urinary sediment and casts decrease, and the dropsy diminish, we may assume that the disease has entered upon a more favorable course. We should now observe attentively the daily proportion of albumin in the urine, and if this diminish coincidently with the improvement in the other symptoms just considered, the outlook will be encouraging. Sometimes, however, after improvement up to a certain point the quantity of albumin continues but little changed, varying only slightly from day to day, and matters may thus continue until the chronic state of the disease is reached.

TREATMENT.

The principles which should govern the treatment of acute nephritis consist : (*a*) in securing so far as is possible physiological rest to the inflamed organs ; (*b*) in limiting the damage progressing in the kidneys, and overcoming the obstruction to the performance of their function ; (*c*) in protecting the system from the consequences of the disease.

In carrying out the first of these indications the patient must be strictly confined to bed, and should lie between woollen blankets. The diet should be regulated in accordance with the intensity of the symptoms. In very acute cases, marked by extreme diminution in the quantity of urine, the latter containing much blood, the diet should be as nearly non-nitrogenous as we can well make it—in accordance with the method of Dr. Aufrecht—and this should be continued for three or four days, or until improvement in the more acute symptoms occurs. The patient should be limited to plain water-gruel made with arrowroot; or perhaps barley-meal; milk and bread are to be prohibited. After the first three or four days it will be necessary to introduce some variety into the diet; and if, as is usual, some improvement in the symptoms has taken place, such articles as turnips, potatoes (especially sweet potatoes), and rice may be added, and the addition of *a little* milk may be made to the gruel.

The bowels should receive due attention. It is the habit with some to begin the treatment with a mercurial purge. It will be remembered, however, that the renal or emulgent veins do not, like those of the other abdominal viscera, empty into the portal system, but directly into the vena cava inferior. Therefore, the venous stasis in the kidneys is not relieved by the action of the mercurial on the liver, as in venous congestion of other abdominal organs. I much prefer a saline cathartic in these cases, as it lowers the increased arterial pressure which is usually present during the first few days of the disease. I quite agree with Dr. Ralfe that these patients should not be overpurged, especially by hydragogues, in the acute stage of the disease; at the same time the bowels should be maintained in a soluble condition. For this purpose a small glass of Hunyadi Janos water may be given in the morning as required, or the Rochelle salt, ℨiv to ℨj—or some such saline may be administered in half a glassful of water.

The question of the use of diuretics is an important one in the treatment of this disease, especially so in the very acute stage. In considering the pathology of this disease in the early part of this section, it was shown that when death resulted from anuria, the kidneys were so intensely congested that numerous extravasations of blood took place in and between the renal tubules. In addition to this, the convoluted tubes were filled and distended by cellular débris and colloid exudation, which prevented excretion of urine. In such a condition of the kidneys, associated as it usually is with hæmaturia, digitalis is one of the most appropriate and efficient diuretics at our command. By increasing the power of the heart, it raises the blood pressure within the vessels of the tufts, thereby facilitating diuresis. By its vaso-constrictor action on the arterioles, it renders their walls less pervious to blood, and thereby checks renal hemorrhage. Therefore, digitalis should always form a part of the diuretic mixtures employed in cases of acute renal inflammation.

As I have recently shown (*Journal of the American Medical Association*, Sept. 12, 1885), renal casts being mostly of an albuminous nature, are soluble in alkaline solutions, as is shown by the facts that they soon disappear from alkaline solutions outside of the body, and that casts are rarely met with in alkaline urine. Now it is well known that when the salts of potassium and sodium formed by combination with vegetable acids are administered, they become changed into alkaline carbonates in their passage through the organism; and as they are mostly eliminated by the kidneys they render the urine alkaline. It is, therefore, easy to render the urine itself a solvent of these plugs, and thus effect their removal, and by so doing we reëstablish diuresis.

That these agents—especially the potassium salts—possess decided therapeutical properties in such cases is shown by the fact that they have won their way into popular favor despite theoretical prejudices against their use. Dr. Roberts (op. cit.) says, "in a considerable number of cases of acute Bright's disease, coming under treatment early, I have obtained almost invariably the best results by the free administration of citrate of potassium. And in no instance, *where the urine has been rendered alkaline* in the first week of the complaint, have I observed the more severe uræmic symptoms, or secondary inflammations."

In the treatment of acute nephritis, therefore, no time should be lost in rendering the urine alkaline, and maintaining it so by the administration of a saline, preferably the acetate or the citrate of potassium. The former should be given in doses of ten grains to ℥ss, and the latter in doses of ℈j to ℥j, both largely diluted with water, and repeated every two to five hours, according to effect on the urine.

If the case be seen at the beginning, it is easier to prevent clogging in the renal tubules and resulting anuria by rendering the urine alkaline, than to relieve the obstruction after it has existed some time. By maintaining the urine alkaline, therefore, from the start, we shall almost invariably avoid anuria, and consequently the more threatening symptoms of the disease.

If the case has attained considerable headway before coming under observation, and the symptoms are acute and threatening, we may resort to measures designed to act more directly on the renal congestion, for more or less time is required to secure the beneficial effects of the alkaline treatment. Among the most valuable of the former measures is leeching. From eight to ten leeches may be applied to the lumbar region, a measure which is often followed by striking and immediate improvement in those symptoms referable to the kidneys themselves. Dry cupping in the same location is likewise, though to a less extent, serviceable in such cases. In all cases in which the symptoms are acute, hot fomentations applied to the lumbar region are beneficial. Flaxseed meal poultices answer the purpose very well, to which may be added sufficient mustard to maintain hyperæmia of the skin.

In addition to these measures, the skin should be stimulated into more or less activity, according to circumstances. For this purpose, as well as with a view to secure cutaneous hyperæmia—a very important consideration in these cases—nothing succeeds so well as the *dry* hot air bath. This may be employed every day, or every alternate day, as described elsewhere, aided by warm drinks, etc. It is only when anuria is marked, and uræmia is threatening, or with a view to relieve the more active dropsical symptoms, that the hot air bath is to be pushed to the fullest extent. Ordinarily a brief impression on the skin of fifteen or twenty minutes' duration—starting the perspiration freely—will be sufficient.

The subcutaneous use of hydrochlorate of pilocarpine is sometimes practised for the purpose of relieving the renal congestion. A case has been reported recently as successfully treated by this means exclusively, by Dr. J. Lazarus (*Wiener med. Presse*, August 2, 1885). The drug paralyzes the vaso-motor system, and the resulting dilatation of the arterioles very decidedly lessens the renal congestion. Pilocarpine, however, should be used with caution if uræmic symptoms be threatening, owing to its tendency to induce œdema of the lungs. The subcutaneous dose of hydrochlorate of pilocarpine is one-eighth to one-quarter of a grain.

The nitrites, more especially nitroglycerine and the nitrites of sodium and potassium, have been employed recently in cases of acute nephritis with apparently good results. The benefits derived from these agents are chiefly due to their power of lowering arterial tension. The nitrites are, therefore, most useful in those cases characterized by exalted tension in the arterial system.

If complications arise during the course of the disease, these will require special measures for their relief. Thus, should acute uræmia occur, which is not uncommon if anuria becomes marked, the eliminative treatment should be pushed more actively. The bowels should be acted upon more freely; diaphoresis by means of the hot air bath or wet pack should be more thoroughly, as well as more frequently employed; in short, the treatment laid down for the relief of acute uræmia, as described in the last chapter, should be energetically employed according to the urgency of the symptoms.

The secondary inflammations within the chest constitute exceedingly dangerous and intractable complications. Arising from the same causes as uræmia, eliminative measures should be employed, as in the latter. In addition to these, counter-irritants (except cantharides and turpentine) and hot fomentations may be applied to the chest. If the pulmonary tissue become invaded, digitalis and ergot combined will often prove a valuable addition to the treatment, especially if begun early. In cases of purulent effusion into the pleural cavity we must resort to quinine, muriate of ammonium, and stimulants. The recent advances in "surgery of the chest" justly lead us to hope that many of such cases may yet be saved by resort to the radical operation.

More frequently, however, if the measures considered have been intelligently employed, the case will assume a favorable course after a few days. The urine will increase in quantity; hæmaturia will subside; pyrexial symptoms will pass away; the quantity of albumin in the urine will diminish; and dropsy will begin to decline.

The most important consideration in the management of the case will now consist in carefully guarding the patient against all causes of relapse, as the latter may occur upon slight provocation. The diet should still be carefully regulated, and meats especially must be prohibited. Starchy and farinaceous articles of diet may be sparingly indulged in, preferably those containing the least amount of nitrogen, such as parsnip, turnip, sweet potato, and thin gruels made with a small quantity of milk, together with such fruits as oranges, pears, grapes, etc. The patient should not be permitted to get up until dropsy has subsided, and the urine has almost reached the normal standard. When the acute symptoms are well on the decline, the patient may be placed on an absolute milk diet; and, in some cases, this measure often effects marked and rapid improvement; but care should be exercised not to resort to the "milk cure" too early. An occasional hot air bath is an excellent measure for the purpose of guarding against accessions to the disease. Diuretics should be avoided in the subacute stage; more especially scoparium, squill, juniper, and such other drugs as stimulate the renal epithelium, or act as vaso-dilators on the renal arterioles. In the declining period of the disease, after all the acute symptoms have been subdued, iron often acts most beneficially, causing a rapid disappearance of the albuminuria, and correcting the anæmia which has resulted from the disease.

Finally, the urine must be examined daily when the disease is under control, and the patient must not be permitted to go about until albumin has completely disappeared from the urine, and even then the urine should be occasionally examined for a period of several weeks to insure the permanent absence of albumin; and only in the latter case should the patient resume his usual diet; indeed, it will be safer for the patient to indulge but sparingly in meats for some weeks after complete disappearance of all traces of albumin from his urine.

As soon as the patient gets up and about, he should be clothed with extra warm woollen garments next the skin; he should go out only in dry and fair weather at first, and he should especially avoid the chilly and damp night atmosphere. If it be convenient to spend a few weeks in a warm climate during convalescence, much benefit will often be derived therefrom.

CHAPTER IV.

CHRONIC NEPHRITIS.

THIS disease has been described under several names, viz.: chronic tubal nephritis, chronic diffuse nephritis, chronic catarrhal nephritis, chronic parenchymatous nephritis, etc. It has been most generally known, however, under the name of chronic parenchymatous nephritis.

The reasons urged against the use of restrictive terms when dealing with acute nephritis, apply with equal force to the disease under consideration; and, therefore, the latter will be dealt with in these pages only under the name of chronic nephritis, which relates in a general though more strictly correct manner to the nature and type of the disease.

ETIOLOGY.

This disease is prone to arise rather later in life than acute nephritis, being most common between the ages of twenty-five and forty years. In a number of cases the disease is a continuation of acute nephritis, and this sequence is said to be most common when the disease arises from exposure to cold.

Scarlatina.—In a considerable number of cases the nephritis consequent on scarlatina passes into the chronic form. This may result from the kidney affection having been overlooked during the course of scarlet fever, or, what is more common, from neglect to pursue treatment to the complete disappearance of all traces of albumin from the urine, in consequence of which the disease lingers without outward manifestations for months. That nephritis of scarlatina may linger in this latent form sometimes very long, is well illustrated in a case recorded by Goodhart (*Guy's Hospital Reports*, vol. xlii. page 147, 1884), in which, after the subsidence of dropsy, etc., following scarlatina, no marked symptoms manifested themselves until six months before death, which did not occur till six years after the scarlatina, when the kidneys were found extremely contracted; showing that the shrinking must have been going on for years.

Pregnancy.—The nephritis of pregnancy, in the vast majority of cases, subsides completely shortly after the termination of gestation. The recurrence of nephritis with successive gestations increases the liability of the disease to pass into the chronic form. In a case of the nephritis of pregnancy which passes into the chronic form, it is altogether likely that some predisposing cause existed before the occurrence of pregnancy; for the disease is sometimes met with in primiparæ.

Chronic nephritis may arise without a preceding acute attack. In such cases it usually develops slowly and insidiously, several weeks or months elapsing before prominent symptoms lead to its discovery.

Cold.—Long-continued exposure to a cold, moist atmosphere is a commonly recognized cause of the disease. Dr. Ralfe believes (op. cit.) that telluric conditions of cold and damp are more powerful factors than the same condition of atmosphere or climate unassociated with influences of soil. In support of this he mentions the fact that sailors, although much exposed to cold and damp, are not particularly liable to chronic nephritis. The disease is certainly most often observed in those whose lives bring them most in contact with cold and damp subsoil influences. Workmen employed in excavating in damp locations, or persons living in damp, partly underground residences, as in basements, etc., are most liable to the disease from the cause under consideration.

Chronic Suppuration.— Necrosis of bone, chronic phthisis, abscesses, and chronic ulcerative processes give rise to chronic nephritis as well as to·lardaceous disease of the kidneys. Bartels believes (op. cit.) that in these cases some product of the suppurative process is taken up into the blood by absorption, and is excreted by the kidneys, exciting inflammation in the latter. It is not known why chronic suppuration induces nephritis in one case, and lardaceous disease in another, though the tendency is evidently greater toward nephritis in adults, while in children lardaceous disease is the more common result. It is not uncommon to find nephritis and lardaceous disease of the kidneys associated as a result of some suppurative process.

Malaria.—Rosenstein, Kelsch, and others have asserted that malaria frequently gives rise to the disease in Europe; and Prof. Atkinson (*Amer. Journ. Med. Sciences*, July, 1884) believes that it is a common cause of the disease in our own country. I have never traced a case to this cause, though malaria is very prevalent in this region.

Syphilis.—Syphilis not infrequently gives rise to chronic nephritis; this may happen quite early in its course as well as during its advanced stage. Dr. Negell (*Journ. of Cutaneous and Venereal Diseases*) points out that syphilitic albuminuria is generally persistent and long continued. Few symptoms being present save albuminuria, specific nephritis may escape notice until œdema appears, and the latter may occur so late that the patient may have passed from under specific treatment, and another cause may be assigned for the nephritis. Again, these patients may be greatly improved under specific treatment, and, on discontinuing the latter the nephritis may linger in a latent form for a considerable time, until at length on coming under the care of the physician again suspicion may be directed to other causes than syphilis. A history of syphilis should be carefully sought for in all cases, for when ascertained, specific treatment often speedily relieves the most threatening symptoms of chronic nephritis.

Alcohol.—The habitual use of spirits is without doubt a cause of chronic nephritis. This may be ranked, perhaps more properly, with the predisposing, than with the exciting cause of the disease.

Finally, a number of cases of chronic nephritis are met with which cannot be clearly traced to any specific cause. It is thought that they owe their origin to morbid conditions of the blood, the nature of which is not well understood.

MORBID ANATOMY.

The macroscopic as well as the microscopic appearance of the kidneys in chronic nephritis will be found to vary widely, according to the stage of the disease in which death occurs. This variation is usually greatest in the two extremes of the disease in reference to the length of time it has been in progress. It will be necessary, therefore, in order to comprehend the whole pathology of the disease, to consider these two extremes separately.

In the earlier stages of the disease the kidney presents appearances similar in some respects to those in the acute form; for, as has already been stated, the chronic is frequently a continuation of the acute form. The kidneys

MORBID ANATOMY. 113

are often enlarged to two or three times their normal size, and their surface is smooth and of a pale yellowish, mottled appearance. The capsule is non-adherent and is easily removed. On section, the capsule retracts, and the cortex gaps, the cut surface presenting the same mottled appearance as the surface of the organ.

FIG. 4.

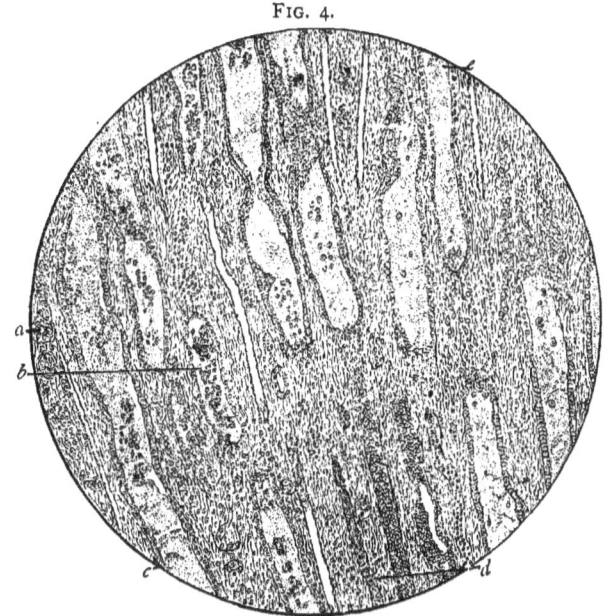

Chronic nephritis; longitudinal section of medulla near papilla. *a.* Contracted tube seen in cross-section. *b, c,* and *e.* Straight tubes containing casts. *d.* Vessel filled with corpuscles.

The yellow spots and streaks observed are swollen tubules filled with fatty material. The cortex is seen to be much thickened; instead of being relatively to the medulla as one to three as in the normal, it is as one to two, or even equal in thickness. The pink cones form a sharp contrast with the pale color of the cortex.

8

On microscopic examination, the tubules are seen to be much swollen and their lining epithelium in a state of marked fatty degeneration. In some cases the renal epithelium is so greatly swollen as to occlude the convoluted tubes. In other locations, tubes may be seen lined with new flattened cells, which are neither fatty nor granular, as in the case of old cells.

The Malpighian bodies may be slightly enlarged, more frequently they are normal in size. Within the capsules of Bowman may be observed fatty degeneration of the epithelium, more especially of that lining the capsule. More or less infiltration with leucocytes is observed both within and without the capsule of Bowman, and to a less extent between the tubules. The straight tubes are often found filled with yellowish colloid material, as are the lower portions of the convoluted tubes immediately above Henle's loops.

The pathological condition of the kidneys resulting from chronic nephritis which will next be described, is always the result of extreme chronicity of the disease. It constitutes the extreme result of the destructive changes after the disease has passed through most of its course. In such cases the kidneys are not enlarged, but often considerably smaller than normal. The organs are firm and dense in consistence, and their surface pale and mottled as before. Sometimes, however, the surface is uneven and granular, but *not red*, as in cirrhosis. The capsule may be slightly adherent at points, and occasionally small cysts may be found upon the surface, or within the cortex, though these are comparatively rare. The cortex is more or less contracted instead of thickened, as it was before.

On examining a thin section of the cortex under the microscope, patches of dense homogeneous tissue are observed, more especially immediately beneath the capsule,

MORBID ANATOMY. 115

and in the course of the interlobular arteries. If examined under a high power, these patches are seen to consist of contracted convoluted tubes, lined with fatty epithelium, Malpighian bodies more or less atrophied; and between these, large numbers of round cells, or fibro-blasts conforming to new connective tissue. Between these solid-

FIG. 5.

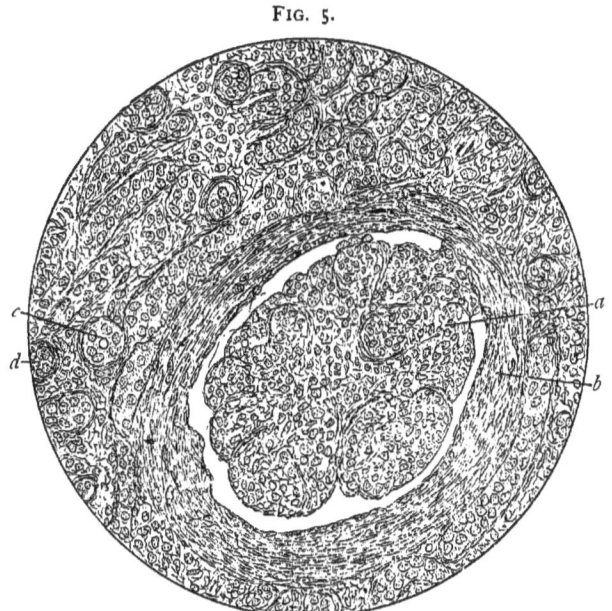

Chronic nephritis; section of cortex. *a*. Malpighian tuft somewhat contracted. *b*. Thickened capsule of Bowman. *c*. Tubule somewhat compressed. *d*. Tubule greatly compressed, with lumen obliterated.

looking patches the glandular tissue is comparatively normal in appearance, save that the epithelium of the convoluted tubes is fatty, and the spaces between the tubules is widened. The capsule of Bowman is usually found thickened, and in some cases the tuft is atrophied. There is considerable infiltration of connective tissue corpuscles in and around the capsule of Bowman, as well as along the course of the interlobular arteries. The muscular coat of the small arteries is thickened, as is also their fibrous tunic.

Such are the principal changes found in the two extremes of this disease. They are so dissimilar that many do not believe that they are the result of the same disease. I hold that the process is essentially the same in both, differing only in degree, and in duration. All the elements of atrophic changes are present in the large kidney, and require only the addition of time to develop them. The tubular changes, always the most rapid in developing, are also the quickest in running their course; while the changes in the connective tissue of the gland are tardy in development, but more pronounced the longer they continue. We may, therefore, find any intermediate form of structural lesion between the two morbid conditions thus described, depending upon the length of time the disease has been in progress when death occurred. To speak of one as interstitial, and the other as parenchymatous inflammation, is incorrect and misleading; the two are inseparable, and one never exists without at least the elements of the other being also present. If one seemingly predominates, it only proves that certain portions of the kidney resist the diseased action a greater length of time than others; and not that one or the other texture of the organ escapes the effects of the disease. The nature of the exciting cause may influence largely the rapidity of the changes, which usually are most marked during the late stages of the disease. Thus, for instance, in scarlatina, the poison or irritant which excites the nephritis, sometimes acts with so much intensity that the connective tissue of the kidney suffers comparatively early from certain pathological changes which are more tardy in development when the disease arises from other and less active causes. I agree with Professor T. Grainger Stewart, in believing that the ultimate tendency of nephritis, from whatever cause it arises, is toward atrophy and contraction of the kidneys, although in some cases atrophy is not extreme, even in cases of long duration.

SYMPTOMS.

The symptoms of chronic nephritis vary much in individual cases, perhaps more so than those of any other form of renal disease. This is due to the facts (*a*) that the disease pursues a more active and rapid course in some cases than in others, and (*b*) that the results of the anatomical changes in the kidneys vary widely in the early and late stages of the disease.

The Urine.— During the early course of the disease the urine is reduced in quantity, its color is usually dark, its specific gravity is more or less above normal, and it contains considerable sediment, which is made up of cellular débris, corpuscles, and casts. The casts vary with the stage and activity of the disease. In the early course, especially if the disease has not been preceded by acute nephritis, the casts may be mostly hyaline, later the epithelial and bloody variety may be observed, while still later in the disease fatty and granular casts predominate. The reaction of the urine is usually acid throughout. The solids of the urine are decreased during the whole course of the disease; urea suffering the most marked reduction, the chlorides less, the phosphates and sulphates least. The urine contains albumin, and, as a rule, in comparatively large quantity. It may, in some cases, reach the proportion of four or even five per cent.;

although it more often ranges near one per cent. The quantity of albumin present is usually greatest during the most active period of the disease. In the early stage of the disease, if it has been preceded by acute nephritis, the quantity of albumin is usually large, and so long as the disease continues unchecked the amount is considerable, although it often varies from day to day. In cases in which the disease arises in the so-called spontaneous manner—*i. e.*, without preceding acute symptoms—the quantity of albumin present may at first be comparatively small. In such cases several weeks, or even months, may elapse before the usual profuse albuminuria is reached.

In the late stage of the disease the character of the urine undergoes considerable changes. The quantity becomes increased to or above normal; the color becomes lighter; the specific gravity descends below normal, and may reach 1.015, or even 1.010. The quantity of albumin is lessened, and the reduction in the quantity of urea now becomes more marked and uniform. These characters of the urine indicate that atrophy of the kidneys has begun.

Dropsy.—This symptom varies inversely with the quantity of urine passed, and hence it is usually most marked during the more active stages of the disease. Dropsy may set in early, or it may not arise until several weeks or months have elapsed; but it is rarely, if ever, altogether absent during the whole course of the disease.

If the disease is the result of acute nephritis from scarlatina, pregnancy, or exposure to cold, the dropsy may subside with the acute symptoms, and the patient may go on for months or years without its reappearance; the urine continuing more or less albuminous. More often the dropsy continues, varying in extent with the activity of the disease. Again, in cases which arise insidiously, or without obvious cause, considerable time may elapse before dropsy appears. In cases which have extended over several years, the dropsy which has been present for a long time may pass away as the result of polyuria induced by atrophic changes in the kidneys.

The dropsy of chronic nephritis varies in extent in each case, yet, as a rule, it is more extensive and persistent than in any other renal affection. It begins in the feet usually, and extends uniformly until the whole subcutaneous tissue becomes infiltrated, and one or more of the large serous cavities becomes involved. If the patient be confined to bed, the œdema may appear first in the scrotum, face, or back. It may apparently fluctuate from day to day in the beginning. For instance, if the patient be about on his feet, the latter may be swollen at night, the swelling partially or wholly subsiding by the following morning; or the face and eyelids may be puffy in the morning, the œdema passing off during the day as the patient goes about.

Once the œdema becomes established, even though limited to the extremities, it usually proves the most obstinate symptom of the disease; indeed, it is rarely checked, save temporarily, until it attains an extreme degree, involving the abdominal and perhaps the thoracic cavities. The limbs become enormously swollen, so that it is difficult, if not altogether impossible, for the patient to move about. The integument becomes greatly stretched, resulting sometimes in rupture, permitting the escape of large quantities of serum. In rarer cases patches of the integument may slough, or become gangrenous, as the result of impairment of the vascular supply through pressure. The dropsy usually involves the abdominal cavity (ascites), and may fill it even to great distention of the abdominal walls. The thoracic cavity likewise, though much less frequently, becomes the seat of serous effusion. This may occur within the pleural cavity, into the pulmonary tissue itself, or, more rarely, into the pericardial sac. Œdema of the glottis is rare; when it does occur, it is immediately threatening to life. The larger number of these cases do not present the more extreme degrees or dangerous locations of dropsical effusion, especially if appropriate treatment be instituted. The dropsy frequently lingers in certain locations, as in the scrotum, feet, or legs, without becoming general, or, if so, not to an extreme degree, perhaps for months, or until some chance exposure or careless action of the patient causes an accession to the disease.

The Circulatory System.—The most marked change here is anæmia, which, in the majority of cases, is striking. The appearance of a patient suffering from chronic nephritis is such as to arrest at once the attention of the experienced physician. The skin exhibits that pale, puffy, anæmic appearance which betokens sanguineous waste, and suggests the presence of some process which is undermining the "vital springs of the economy." It is only exceptionally that anæmia is absent in these cases, even in the early stage of the disease. Changes in the structure of the heart or arteries are rare in chronic nephritis, and are confined to very late stages of the disease, when contraction of the kidneys is in progress. The pulse is small, compressible, and accelerated, and its tracing by the sphygmograph is marked by dicrotism. These conditions undergo a change in the late stages of the disease, coincident with atrophy of the kidneys; but even in such cases hemorrhages into the brain or retina are rare, owing, doubtless, to the anæmic condition. Carditis is a comparatively rare secondary complication of this stage.

SYMPTOMS.   121

The Muscular System.—Debility and emaciation are constant and early symptoms of the disease. The emaciation may not be apparent, being masked by dropsy; but if the œdema subsides, or is absent, it is usually quite noticeable, and in some cases extreme. The strength declines to a marked degree tolerably early in the disease, and the patient goes about only with considerable effort, and he is easily fatigued. Sooner or later, physical exercise becomes so exhausting that the patient voluntarily confines himself to the house, and perhaps to bed; asthenia becoming more marked as the disease continues.

The Digestive System.—The appetite becomes impaired early in the disease and usually remains so throughout. Bartels has remarked that these patients possess an especially poor appetite for meats. Nausea and vomiting are not uncommon in chronic nephritis. The matter ejected by the stomach is usually of a watery appearance, and is doubtless the result of œdema of the gastric mucous membrane, as it occurs chiefly during the height of dropsical effusion elsewhere. Purely reflex vomiting occurs less frequently, but when it does occur it is usually either during an acute accession of the disease or during the late stage. Diarrhœa sometimes sets in and greatly adds to the debility of the patient. This is usually due to œdema of the intestinal mucous membrane. Like vomiting, it is most common during the height of the disease. The stools are watery in appearance and contain much serum. In the late stages intestinal ulceration sometimes occurs, giving rise to diarrhœa of a different character, the stools containing pus, blood, mucous shreds, etc.

The Nervous System.—In the early history of the disease adynamia is the most constant characteristic, and this becomes marked as the disease progresses. The more active disturbances of the nervous centres known as uræmic are rare in the early stages, and, if exceptionally they do occur, they are usually the result of the disease approaching more nearly the acute type. Various reasons have been assigned for the comparative rarity of uræmic symptoms in chronic nephritis. Thus, Bartels has remarked that the production of urea in the system is diminished. The appetite and digestion are poor, and especially so for nitrogenous foods, and thus the most prominent source of urea is greatly reduced. The muscular system being much wasted in these cases, tissue changes resulting in the liberation of nitrogenous products in the organism are less active. Again, the almost constant coincidence of dropsy with the more active grades of the disease, serves to assist in protecting the system from the consequences of the defective renal function. The retained urinary derivatives are largely held in solution in the serous fluids stored in the great cavities, or subcutaneous tissues outside of the circulating fluid, where they are unable to reach, and consequently to irritate the nervous centres. We know that urea at least is thus largely stored up, for it has been found copiously in the serum withdrawn from the abdominal cavity, as well as in fluids ejected by the stomach and bowels.

In the smaller number of cases which reach the stage of renal contraction, uræmic symptoms are more common, but even here uræmia is less frequent than in genuine cirrhosis of the kidney. This is probably due largely to the reasons which have just been considered. If amaurosis occurs, it is usually the result of retinal degeneration, the approach being gradual and the impairment of vision permanent. Convulsions and coma, when they occur, nearly always prove fatal.

The secondary inflammations are more common than uræmic disorders. Of these, pneumonia is perhaps the most frequent.

COURSE AND DURATION.—The course of chronic nephritis depends much upon its cause and management. Those cases which result from acute nephritis, if properly treated, sometimes pursue a favorable course. The dropsy subsides after a few weeks; the albumin diminishes and finally disappears completely from the urine; and the patient recovers. More often, however, the symptoms are obstinate and the disease does not yield to treatment. The dropsy may continue for months, or even years; or it may become modified or pass off, albuminuria remaining. The disease may continue without marked change for a long time till some disturbing cause lights up more active changes which may prove fatal, or, these being avoided, the case may continue for years, resulting ultimately in contraction of the kidneys.

In those cases in which the disease follows in the wake of some other process, its course is usually more rapid, as the system is already reduced and incapable of withstanding the effects of the nephritic complication. In cases arising without apparent cause, the early course of the disease may be unattended by dropsy, or other marked symptoms, for a long time. In one of my cases of anæmia with slight albuminuria, for more than two years the sole indication of the existence of the disease, except the albuminuria, was the presence of granular and fatty casts in the urine.

It will thus be seen that the duration of chronic nephritis, like its course, is exceedingly variable. Some cases succumb within a year to extreme dropsy, exhaustion, or some other complication; while others survive for two, five, or even ten years.

DIAGNOSIS.

CONSTRUCTIVE.—In well-marked cases the recognition of this disease is not difficult. The points to be kept prominently in view are as follows: Marked anæmia; general dropsy; copious albuminuria; diminished urine, containing fatty and granular casts, and numerous leucocytes. With these symptoms are usually coupled early debility and impaired appetite.

In less typical cases, however, difficulty may be experienced in recognizing the disease, especially if its approach is insidious, and unaccompanied by dropsy or copious albuminuria for a considerable time. Anæmia, fatty casts and leucocytes in the urine, and early debility are usually present, but it may be necessary to reach a conclusion through exclusive rather than direct symptoms.

DIFFERENTIAL.—Chronic nephritis may be confounded with *cyanotic induration* of the kidney. The distinction is easy, however, if the following points are carefully remembered. In cyanotic kidney albuminuria is slight; the casts are of the hyaline variety; leucocytes are rarely present in the urine; dropsy is chiefly confined to the lower extremities; the general condition of the patient is good; the breathing is embarrassed; the heart is obviously defective; and the appearance of the skin is dark and cyanosed. Most, if not all, of these symptoms form a sharp contrast with those of chronic nephritis.

Lardaceous disease presents many similar points to those of chronic nephritis. The following are the distinctive points to rely upon. In lardaceous disease the urine is constantly above normal in quantity; gastro-intestinal disturbances are more early, and more persistent than in chronic nephritis; especially is this the case with diarrhœa.

The liver, or spleen, or both, are likely to be enlarged in lardaceous disease; and lastly, lardaceous disease is the outgrowth in most cases of some chronic malady to which it may be readily traced. Chronic nephritis and lardaceous disease may coexist, and in such cases it may be impossible to determine which is the original malady. The symptoms of *cirrhosis of the kidneys* closely resemble those of chronic nephritis when the latter has entered upon the stage of contraction. In such cases the history throws important light upon the question. Thus, if the case is one of primary cirrhosis there will usually have been no dropsy, save in the very latest stage of the disease. The urine will have been constantly above normal in quantity, containing but a relatively small amount of albumin, and its specific gravity will have been below normal from the beginning. Cardio-vascular changes occur early and precede the appearance of dropsy, if the latter occurs at all in cirrhosis; while in chronic nephritis, dropsy precedes the cardio-vascular changes, and, moreover, they rarely become so marked as in the case of cirrhosis.

The dissimilarity in the manifestations of cirrhosis, chronic nephritis, and lardaceous disease of the kidney is presented in strong contrast in the differential table in the chapter devoted to cirrhosis of the kidney.

PROGNOSIS.

It must be candidly admitted that the prognosis in this disease is generally grave. While I am not in possession of statistics showing the actual proportion of deaths, yet I have no doubt that the majority of persons who become the subjects of this disease ultimately succumb to it, even under the most skilful management.

While the above is probably strictly true, yet it is by no means meant to imply that treatment is without influence on the disease, or that cases do not sometimes recover. The prognosis in individual cases will be influenced by several circumstances. The length of time the disease has been in progress will influence materially the ultimate result; the more recent its origin, circumstances being equal, the more favorable will be the prognosis. It is true that some cases recover completely after lasting one, two, and even three years, but as the duration of the disease becomes extended, especially beyond the first year, the prognosis becomes increasingly gloomy.

The cause of the disease gives important indications as to the prospects of recovery or otherwise. If it has arisen in consequence of some serious constitutional state which itself presents slight hopes of recovery, the prognosis will be unfavorable. If, on the other hand, the disease be traced to syphilis, and specific treatment be employed, we are justified in expecting a favorable termination. In cases which originate in acute nephritis, as from scarlatina, pregnancy, and exposure to cold, much will depend upon the intensity of the disease and length of time it has existed *in the acute form*. The more intense and long continued the acute changes, the more irreparable the damage resulting to the organs, and consequently the more unfavorable the prognosis.

Certain manifestations of the disease itself may be looked upon as favorable indications, such as disappearance of dropsy, decrease of albumin in the urine, and especially the proportions of urea and solids in the urine approaching and remaining near the normal standard. On the other hand, if the dropsy attain an extreme grade and thus continue; if the urine remain below normal in quantity, and contain much albumin and sediment; and especially if the proportion of urea in the urine remain much reduced, we know that the kidneys are extensively involved in pathological changes, and hopes of recovery become greatly diminished in consequence.

TREATMENT.

In the treatment of chronic nephritis the first point of importance is to ascertain, if possible, its cause, and more especially to determine if it be still operating, and thus keeping alive the disease ; if so, we should proceed at once to remove it so far as it may be possible to do so.

Thus if the system be tainted with syphilis, specific treatment should be energetically employed. Any suppurative process going on in the organism must be met by appropriate surgical measures for its relief. Habits of alcoholism must be corrected. The surroundings of the patient should receive due attention, and if he reside in some damp, partly underground location, a change to a more favorable situation is imperatively required.

The hygienic measures considered in the chapter on albuminuria are eminently appropriate in the management of this disease, and they need not be repeated here. The diet, in most cases, may be more liberal than that permitted in acute nephritis, because (*a*) the general condition of the patient usually demands this, and (*b*) the comparative rarity of uræmic complications in this disease permits a more generous diet without incurring dangers from that source. At the same time, the function of the kidneys must not be unduly taxed by indulgence in large quantities of nitrogneous foods.

We next have to deal with the medicinal and special treatment of the disease.

THE EARLY STAGES.—In the early course of chronic nephritis two classes of cases are met with in which the type of the disease differs essentially, and therefore they require separate consideration.

The first of these comprises those cases which follow upon or are a continuation of acute nephritis. In such cases the disease conforms more closely to the acute type, and therefore the treatment should likewise correspond, in a large measure, to that of acute nephritis. The urine is diminished in quantity, of high specific gravity, and contains much albumin. Dropsy is more or less prominent. The measures advised for similar symptoms in acute nephritis are now indicated, though they do not require to be pushed so actively. The citrate and acetate of potassium should be given in sufficient quantities to maintain the urine alkaline. Mild purgatives should be administered sufficiently often to equalize the circulation and encourage osmosis. Hot air baths may be employed as frequently as the strength of the patient will permit without inducing depression. The patient should be confined to bed so long as dropsy is marked. Meat should be excluded from the diet in this condition.

Under the influence of such measures the renal congestion will usually yield. The quantity of the urine will increase, dropsy will decline, and the quantity of albumin in the urine will decrease. In short, the more active symptoms of the disease will subside; and under appropriate measures, to be considered later, complete recovery will follow. In many cases, however, improvement stops at a certain point, and the disease continues without marked fluctuation, seemingly little influenced by medication.

In the second condition met with in the early course of this disease the symptoms from the beginning are less urgent. The approach of the disease is insidious, and is accompanied by but slight albuminuria. Dropsy if present is not marked, and the urine is nearly normal in quantity. Especially is the use of diuretics and purgatives in this condition to be avoided; they should be reserved to combat complications of the disease when they arise, such as uræmia, dropsy, constipation, etc. We have to deal here with a chronic disease for which we possess no specific remedy. Therefore, only remedies should be employed which are calculated to serve definite purposes. The more active changes in the kidneys have not yet been entered upon. Those existing, consist chiefly of fatty changes in the glandular structure, and these may pass away if only the organs be carefully guarded against *all* causes of irritation, their sources of nutrition improved, and healthy reparative changes encouraged.

In the condition under consideration no single remedy approaches iron in point of efficiency, as its action is naturally directed against the more prominent consequences of the disease, both general and local, viz., anæmia and degenerative changes. Some preparation of iron should, therefore, be given persistently. The best effects of this agent are usually obtained by occasionally substituting one preparation for another. Of these, I prefer the subcarbonate, tincture of perchloride, potassio-tartrate, and citrate of iron. As we possess so many excellent preparations of this metal, doubtless others are as efficient as those named above. An object to be kept in view is the selection of such as will prove least embarrassing to the weakened stomach, while they can be administered in sufficient doses to correct the anæmic condition of the patient Nux vomica or strychnine acts as an efficient tonic in these cases, improving the condition of the stomach as well as increasing the tone of the motor system. Quinine is also serviceable, and it may often be advantageously combined with either nux vomica or some preparation of iron. Aside from the remedies just named, there has been a growing tendency of late to rely more on the hygienic, and less on medicinal treatment in chronic nephritis, unless special symptoms or complications arise requiring attention. In this connection the so-called "milk cure," or the treatment of the disease by a milk diet, may be briefly considered. Of late, much attention has been given to the use of milk in the treatment of nephritis. Among others, Dr. George Johnson advocates an absolute milk diet in these cases. He has informed me that, in his experience, the slightest addition of other food, such as a piece of cracker, or even a little arrowroot, often increases the quantity of albumin in the urine, or sometimes causes it to reappear if used too early. I know from personal observation of cases in charge of this distinguished physician in King's College Hospital, that the results which he obtains from an absolute milk diet are most excellent; often consisting in a complete disappearance of albumin from the urine in from two to six weeks. Subsequent observations have convinced me that the enforcement of an exclusive milk diet proves the most successful measure in a considerable number, though not in all of these cases. Unfortunately, serious drawbacks are encountered in many cases in attempting to enforce the exclusive use of milk, and these may render it impossible to carry out this treatment. Milk does not contain in proper proportions the elements necessary to sustain life. Thus Senator has shown (*Albuminuria in Health and Disease*) that while 2 litres of milk contain about 74 grammes of fat, only about 30 are required for the use of the system. On the other hand, 2 litres of milk contain only about 70 grammes of albumin, and, according to Voit, at least 85 grammes of albumin a day are necessary to support life. Again, 2 litres of milk contain only about 100 grammes of hydrocarbons, while the requirements of the system demand about 300 grammes. It will thus be seen that if we give sufficient milk to supply the demands of the system for albumin and hydrocarbons, we have three or four times the amount of fats necessary, and this overplus of fatty food in many cases disagrees with the stomach. With a view to obviate this, it is found best to use skimmed milk, at least in the beginning. After a time, as the stomach becomes accustomed to the skimmed milk, the latter may be gradually changed to unskimmed milk. In many cases, especially if digestion be vigorous, the patient may be able to subsist for a long time upon unskimmed milk alone. In such cases the extra fat takes the place of the deficient hydrocarbons, the stomach being capable of transforming the former. Caution is necessary in first enforcing a milk diet; it being safer to bring this about gradually, in order to test the capability of the stomach, and to accustom the organ to the change. Taking a part of the milk in the form of curds, as suggested by Dr. Embleton, is often an excellent method, as of all forms of solid food curds are the least liable to increase albuminuria. Koumiss has been advised where milk disagrees, as has also buttermilk, and the latter may be partly substituted in some cases for milk. Peptonizing the milk will render it more digestible, but in view of the fact that cases have been recently reported (*British Medical Journal*) in which the use of peptonized milk induced albuminuria in apparently healthy people, it may be well to avoid it in these cases till further observation shall have determined the risks to be incurred through its use. Boiling renders the milk more acceptable to the stomach in many cases, and this may be safely tried in all cases.

In a number of cases, unfortunately, a milk diet disagrees with the patient, and in such it is useless, or even injurious, to persist in efforts to reconcile the system to its use. In still other cases, while the milk diet agrees with the patient, yet its use is not followed by any marked improvement. In these cases the best course is to supply a diet containing but a small proportion of nitrogen, in accordance with the hygienic principles laid down in the chapter on "albuminuria."

A number of other agents, such as tannate of sodium, fuchsin, ergot, tannic and gallic acids, benzoate of lime, apomorphia, eucalyptus globulus, copaiba, tar, sandalwood oil, squill, corrosive sublimate, calomel, arsenic, phosphoric acid, have been recommended in the state of the disease under consideration. The first four or five of these seem to diminish the quantity of albumin in the urine in many cases, and their use is not attended by any injurious effects, so far as I am aware, yet I have little confidence in their curative influence. The more stimulating preparations, such as the balsams, are of doubtful utility, and should be used only with caution; while the last four named in the list are harmful — in fact, they predispose to fatty changes, which are the predominant lesions in the kidneys in this disease. Their use should, therefore, be avoided. I have had some experience in the use of tannate of sodium and fuchsin, and I have usually found their exhibition to be followed by a decrease in the quantity of albumin in the urine. I have not observed any other favorable change occur in the course of this disease through their use. Still, as a decrease in the quantity of albumin in these cases is itself a desirable object; and, moreover, since the tannate of sodium and fuchsin probably exercise no injurious effects, they may be employed. The tannate of sodium may be administered in doses ranging from five to twenty grains, preferably dissolved in water. The usual dose of fuchsin in these cases is from one to three grains, given in pill. It is said that fuchsin does not restrain the loss of albumin by the kidneys, unless it colors the urine red. Of late, caffeine has been recommended in these cases. Dr. Gubb (*Lancet*, February 28, 1885) reports a case successfully treated by this agent. Gubler has shown clinically that caffeine acts as a diuretic, and it has been used to relieve cardiac dropsy. Brackenridge claims (*Edinburgh Medical Journal*, xxvii. 4, 100) that caffeine is a "stimulant of the renal glandular epithelium." Its diuretic properties have been ascribed by some to its action on the cardiac nerves, while others have attributed them to its direct action upon the nerves of the kidney. Further observations are necessary, however, in order to determine its range of therapeutic application in these cases. The dose of the citrate is from one to five grains, and it is usually given three or four times daily. If it affect the sight or gait, or excite nausea, headache, palpitation, etc., the dose should be decreased.

LATE STAGES.—Notwithstanding the best directed efforts, some of these cases pass on unchecked, until more extensive and serious lesions in the kidneys are the result. Should such be the case, dropsy will reappear, or, if it has not been present before, will now become developed. The quantity of albumin in the urine will increase, and there will be numerous fatty casts. The quantity of urea daily excreted by the kidneys will diminish, as will the quantity of the urine itself. These symptoms are not of sudden onset, as in acute nephritis, but creep on slowly, keeping pace with the advancing lesions in the kidneys. They are usually months or years in attaining their full development. At their height some of them become so threatening that special measures for their relief are called for. This is most often the case with dropsy. It should be met with diaphoretics—preferably the hot air bath—purgatives and diuretics. These measures, however, cannot be pushed to the same extent as in acute nephritis, on account of the greater debility of the patient. An occasional hot air bath may be given, or jaborandi may be administered, if it do not prove too depressing. If the strength of the patient permit, and no diarrhœa be present, an attempt may be made to reduce the dropsy by means of purgatives. The saline purgatives are the most suitable in these cases, given as I have already described.

As a rule, however, it will be best to endeavor to reduce the dropsy by increasing the action of the kidneys; and thus secure relief through the natural channels. Numerous agents have been recommended for this purpose, and though our choice of these is not so restricted as in acute nephritis, yet the extensive degeneration going on in the kidneys demands, even in these cases, that we exercise discrimination in the employment of diuretic medicines. It is probably safer, therefore, to avoid the use of the more stimulating diuretics, especially turpentine, cantharides, juniper, etc.

Owing to the marked debility present, the cardiac force is diminished, and this weakens the functional power of the kidneys. Some agent such as digitalis should, therefore, be combined with the diuretic mixtures employed. In my experience, the fluid extract of adonis vernalis, in ten drop doses, often adds remarkably to the action of digitalis on the heart when given with the latter. As at least one of these medicines is said to be cumulative in its effects; and as it is usually necessary in these cases to stimulate the cardiac power directly and continuously, for a considerable length of time, convallaria majallis may be substituted occasionally for either or both of them. Convallaria majallis will usually be found less effectual than digitalis, though in many cases it increases the cardiac power to a marked degree. Among those agents that act as diuretics directly on the kidneys, scoparius (broom) has long enjoyed a high reputation in these cases. It is best given in the form of infusion, which may be prepared by steeping an ounce of the tops in a pint of water for half an hour. Of this, from one to two fluidounces may be given four times a day. In large doses it is cathartic.

Apocynum cannabinum often acts as an efficient diuretic in these cases. I have known it to stimulate enormously the secretion of urine. Unfortunately, however, in the doses requisite to obtain its more active diuretic effects, it often produces sharp purgation and even vomiting. The dose of the fluid extract is from five to twenty drops. It may be administered in the form of infusion; its preparation and dose are similar to those of scoparius.

When the morbid anatomy of this disease was under consideration, it was shown that the renal tubules were often found filled with colloid matter, more especially the straight tubes. This condition may become so marked as to interfere at times with the excretion of urine. It has also been shown that when the urine is rendered alkaline the free colloid matter becomes dissolved and is washed out with the urine. It is well, therefore, when the urine contains many casts, to render it alkaline and to maintain it so for some time by the administration of citrate or acetate of potassium. This measure of itself will sometimes reëstablish diuresis. Immermann insisted long ago (*Correspondenz Blatt für Schweizer Aerzte*, June, 1873, No. 11) that the use of acetate of potassium in these cases often relieved the dropsy and other symptoms with astonishing rapidity.

Sometimes when distention of the skin becomes very great from underlying œdema, relief may be afforded by means of punctures; but care should be observed not to make extensive incisions, for large wounds not only may refuse to heal under such circumstances, but also slough extensively, or even become gangrenous. These punctures should not be numerous, or situated near each other, for the same reasons. Dr. Ralfe (op. cit.) advises the use of Southey's fine canula for this purpose, one puncture, or not more than one for each leg, being required. In all cases, wounds made for the purpose of relieving œdematous distention of the parts, should subsequently receive careful surgical treatment until they are healed, with a view to prevent the bad results which are liable to follow.

Excessive accumulation of serous fluids in the great cavities of the abdomen and chest may be relieved temporarily by means of the aspirator.

In those rare cases in which œdema invades the glottis, various measures have been suggested for its relief, but prompt punctures or incisions are the most efficient means of relieving the patient and saving his life.

A limited number of cases of chronic nephritis survive the consequences of the disease, which continues unchecked till the kidneys at last become the seat of atrophic changes. Oliguria now gradually becomes replaced by polyuria, and as the latter becomes established dropsy, if previously present, passes away, or becomes modified. Although the absolute quantity of the urine becomes increased, the proportion of its contained solids, especially that of urea, undergoes no augmentation; in fact, the contrary is usually the case; and thus the patient is subjected to increasing risks of uræmia and secondary complications. Cardio-vascular changes, similar to those in true cirrhosis, may follow, though usually less extensive than in the latter disease.

The treatment of this stage of the disease should be conducted on the same principles as in cirrhosis of the kidney, the only essential difference to be kept in view is the greater anæmia and debility of the patient.

The nitrites will be useful if the pulse presents evidences of high tension, otherwise not; in fact, here, as in cirrhosis of the kidney, the nitrites, such as nitroglycerin, nitrite of sodium, etc., affect the disease through their power of lowering high arterial tension.

I have obtained good results from the administration of chloride of gold in some cases of chronic nephritis. It occurred to me some time ago that if chloride of gold be capable, as has been claimed, of checking the growth of new connective tissue in the kidneys in cases of cirrhosis, it should be still more capable of *preventing* such growth in cases of chronic nephritis. Acting upon this thought, I have administered the chloride of gold in the late stages of chronic nephritis in a number of cases, with good results. The most recent case of the kind is that of a patient whom I have now had under treatment over two years. The acute symptoms were soon subdued, including dropsy and some uræmic disturbances. For the past twenty months, however, the disease has been apparently stationary, or nearly so, up to within the past three months, although the most rigid hygienic measures have been faithfully carried out, and the medicinal treatment carefully regulated to the requirements of the case. The urine all this time has contained a considerable amount of albumin, and anæmia has been marked throughout, although iron has been given almost continuously. Cardio-vascular changes are absent, as will be seen from the sphygmographic tracing below, which is essentially the same as several taken within the past year.

FIG. 6.

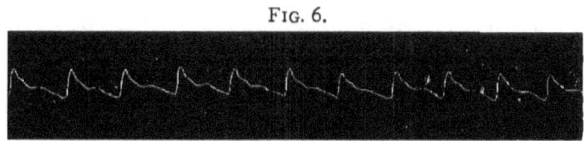

Three months ago this patient was put upon chloride of gold and sodium, in one-thirtieth of a grain doses, repeated three times a day, and improvement became apparent in about two weeks, which has continued uninterruptedly to date. The quantity of albumin in the urine has steadily decreased, the strength and color of the patient have uniformly improved under increasing doses of the chloride of gold, which is now being taken in one-twentieth of a grain doses. I copy from my notes, November 16, 1885, the following: Mrs. ——, much improved in appearance, cheeks ruddy, etc.; urine three pints, normal in color, reaction acid, specific gravity 1.018, albumin present in faint trace, only discoverable by means of delicate testing. The patient expresses herself as feeling altogether up to her usual standard of health and strength during the past month, and this for the first time in the past two years.

Whether chloride of gold acts more efficiently in checking the growth of new connective tissue in the kidneys in chronic nephritis, than in cases of renal cirrhosis, or whether it is because we are able to anticipate these changes earlier in the former that gold there proves most efficacious, I am unable to say; but I am satisfied that the use of this drug is followed by more marked improvement in the late stages of chronic nephritis than in genuine cirrhosis of the kidneys. Aside from the above, I have nothing special to add as to the treatment of chronic nephritis when it reaches the stage of contraction. Both the disease and its complications should be treated substantially as in cases of true cirrhosis of the kidneys, which will be considered at length in the following chapter.

CHAPTER V.

CIRRHOSIS OF THE KIDNEY.

This disease has been described by different authors under various names, viz.: chronic interstitial nephritis, granular degeneration, granular kidney, granular atrophy, gouty kidney, chronic Bright's disease, renal cirrhosis, etc. The name cirrhosis, first applied to the disease by Professor Grainger Stewart, seems, on the whole, the least· open to objection.

ETIOLOGY.

Among the alleged causes of this disease may be mentioned, age, habits of the patient, heredity, psychical influences, gout, chronic dyspepsia, malaria, saturnine influences, climate, and chronic diseases of the lower genitourinary passages, especially those of an obstructive nature.

Age.—Among the predisposing causes of cirrhosis of the kidney age seems to occupy a prominent place. The disease occurs but rarely in childhood, more frequently in young adults, most frequently in middle and advancing life, and less frequently again in extreme old age. Dr. Dickinson collected from the *post-mortem* books of St. George's Hospital the records of death in 242 cases, and of these the highest mortality occurred between the ages of forty and fifty years, the second highest mortality was between fifty and sixty years, and the third highest was between thirty and forty years. The statistics of other observers substantially confirm the figures of Dr. Dickinson.

It is only fair to consider, however, in the first place, that the figures given refer merely to the periods of mortality of the disease; and in the second place, that the duration of this disease is to be measured "by years rather than by months;" it being not uncommon for it to exist for five or ten years before death. It is not unlikely, therefore, if we could fix upon the precise beginning of the disease in all cases, that it would be found to arise more frequently at an earlier period than is now generally supposed.

Cirrhosis of the kidney occurs three or four times more frequently in the male than in the female sex, which does not seem at all singular if we reflect that the male sex is much more exposed to the causes of the disease than the female.

Habits.—This disease is especially prone to arise in people who are what has been termed "*high livers;*" that is to say, those who indulge very largely in meat diet; especially the highly seasoned meats, and those rendered more palatable by the addition of rich gravies. Such foods leave a large amount of nitrogenous waste behind, which it is the office of the kidneys to eliminate from the system; hence the ingestion of large quantities of such food unduly taxes the kidneys, leading to functional, and at length to structural impairment of the organs. This cause becomes still more active in persons who are in the habit of using fluids sparingly. As a result of this, the urine, already highly charged with solids (the product of increased functional activity), becomes still more concentrated, and its solid elements, when feebly diluted, act as irritants to the renal structures. Again, if the urine becomes very concentrated, some of its solids—notably the urates—are liable to fall out of solution, and in such cases they always prove more or less irritating to the kidneys and urinary passages. If we examine the urine in the *very earliest stages* of the disease, we find it very concentrated, and depositing crystals of uric acid and calcium oxalate.

There is much difference of opinion as to the causative influence of alcoholic drinks in this disease. Bartels holds that, at least in Germany, it cannot be ranked among the causes of the disease. On the other hand, Grainger Stewart believes that intemperance is a very common cause of the disease in Great Britain. This discrepancy of opinion may arise from the fact that, in Germany, alcoholic beverages are generally used in a largely diluted form, as the lighter wines and beer; while in the British Isles the use of spirits is more largely indulged in. It has already been shown that not only albuminuria, but also nephritis may be produced in animals by administering to them alcohol, and, therefore, it is not unreasonable to suppose that when some predisposing tendency to the disease exists, the liberal indulgence in alcoholic beverages, especially in concentrated form, may act as an exciting cause.

Psychic Influences.—While these are not well understood in their relation to this disease, yet the opinion has become prevalent of late that mental influences of a decided and continuous character tend to favor the production of this disease. The nervous depression consequent upon prolonged grief, or business anxiety, has been assigned as the cause of not a few cases. Indeed, in a record of thirty-two cases, one author claims to have traced the cause to such influences in three-fourths of them. It is extremely difficult to fix the cause definitely in such cases from their very nature; and thus far we are obliged to depend upon inference, rather than proof.

Gout.—There can be no question that cirrhosis of the kidney is very frequently associated with gout, as Dr. Todd first pointed out; and it is believed by many to be the cause of the cirrhosis in a large number of cases. This belief is strengthened by the fact that in such cases gout often precedes for a long time the appearance of cirrhosis. Dickinson mentions a case in which gout existed "off and on for twenty-six years," and which subsequently terminated fatally as a result of cirrhosis of the kidney. On the contrary, others (among them Virchow) think that the renal is the primary affection, and that gout is the result thereof, being brought about by failure of the kidneys to excrete the uric acid from the blood. Whatever may be the precise relationship existing between gout and cirrhosis of the kidney, there is no doubt of their frequent association.

Dyspepsia.—This has been supposed by several authors to be one of the causes of renal cirrhosis. Dr. George Johnson was the first to advocate this view. He holds that the "renal degeneration is brought about by the long-continued elimination by the kidneys of the products of faulty digestion." This view has been to some extent accepted in England, while in Germany it has been dissented from. That dyspeptic disorders accompany the early stages of renal cirrhosis, is a fact of common observation; indeed, it may be regarded as one of the most constant of the early manifestations of the disease. In my own experience, however, I have been able in every instance to determine from other evidences that the disease was in actual progress at the time the dyspeptic symptoms first manifested themselves; and hence I have regarded the gastric disturbances as the result, rather than the cause of the disease in such cases.

In four cases I had an intimate knowledge of the condition of the patients for several years prior to the appearance of renal cirrhosis, having been their medical attendant. In three of these cases, while dyspepsia was among the early symptoms, yet other indications of the disease were present quite as early—such as high vascular tension, and occasional albuminuria—which showed that the disease must have been in progress before the development of dyspepsia. In the fourth case—still under observation—the patient has *never* suffered from any dyspeptic symptoms to date, though over a year ago his pulse showed marked high tension; later on, the urine became light in color, of low specific gravity, and occasionally albuminous, which, with obstinate epistaxis, and quite recently an attack of gout, leave no question whatever but that the case is one of genuine cirrhosis of the kidneys. Again, the dyspeptic symptoms common in early cirrhosis are uniformly benefited by a course of diaphoretics, cathartics, and other eliminative measures, and the inference is very strong that the cause of the gastric disturbance is nothing more than the irritation set up by the retained urinary elements in the organism —in short, that it is uræmic.

In my experience, if the quantity of the urine is maintained well up to the normal, the dyspeptic symptoms not only pass away, but the appetite returns, and the patient increases in flesh, where before he had been losing weight. For these reasons I look upon the dyspeptic symptoms of renal cirrhosis as a result, rather than a cause of the latter.

Heredity.—Renal cirrhosis owes its origin in some cases without doubt to hereditary influences. In some families this tendency is scarcely less marked than is that of tuberculous disease. Drs. Dickinson and Tyson have both recorded remarkable instances of this kind, in which numerous members of the same family have succumbed to the disease through several generations. Similar though less marked instances have been noted by other writers; and a minute inquiry into the family history of these patients will very often elicit evidences of the previous existence of the disease, either immediate or remote, or both.

Plumbism.—The frequent appearance of renal cirrhosis in those who have been long subjected to the influence of lead, leaves little doubt that the latter is a common cause of the disease. Dr. Garrod has shown that gout is frequently induced by plumbic influences, and some observers think that the renal disease is the result of the gouty influence in these cases, rather than directly due to the lead. This does not alter the fact that renal cirrhosis follows in a very large number of those exposed to continuous plumbic influences. Dr. Dickinson's statistics, compiled from St. George's Hospital reports, show that of forty-two workmen whose occupations brought them in daily contact with lead in some form, about two-thirds eventually suffered from granular kidneys.

Climate, Malaria, etc.—The geographical distribution of this disease would seem to show that its production is influenced more or less by physical surroundings. It is most common in temperate climates, and least so in the border ranges between temperate and torrid zones. All renal diseases are influenced by vicissitudes of temperature and humidity of the atmosphere, and these are found most injurious in the cooler ranges, such as the temperate zone, between the fortieth and sixtieth parallels. Stricture of the urethra, and catarrhal inflammations of the urinary tract, especially of the renal pelvis, are looked upon as not infrequent causes of this disease. Of these affections, calculous pyelitis seems to be the most frequently associated with renal cirrhosis. The only other influence that I am inclined to ascribe as a cause of this disease, though perhaps an infrequent one, is that of malaria. I have observed cirrhotic disease of the kidneys in a few cases in which no cause seemed apparent except malaria. One case recently came under my observation, that of a surgeon in the American civil war, who became so much invalided through inveterate malarial poison, that he was obliged to abandon his practice in Indiana and remove to the borders of our lake, with the hope of throwing off the malarial influences; but only with partial success. He succumbed at length to well-marked renal cirrhosis, which in his own as well as my belief, was due to the malarial poison.

MORBID ANATOMY.

In typical cases of renal cirrhosis, the kidneys are found after death more or less diminished in size, according to the length of time the disease has been in progress. I have seen the kidneys so contracted that both organs together weighed less than one normal organ. This contraction is seldom symmetrical, one kidney usually differing in size more or less from its fellow. An abnormal quantity of adipose tissue usually surrounds the glands, and is deposited in the pelvis of the organs, around the calices. The surface of the organs is of a reddish-brown color, mottled over with yellow spots about the size of pin-heads; and it also presents an uneven appearance, due to small granular elevations which everywhere cover the external cortex. These little granulations vary in size from that of a millet-seed to three or four times larger. Between them may often be observed small cysts, filled with yellowish translucent fluid. These cysts are usually several times larger than the granular elevations surrounding them; occasionally they are as large as a marble. The capsule is thick, and adherent to the cortex beneath, which latter clings in places to the capsule if stripped from the surface. The organ feels firm and resistant to the fingers, not unlike soaked leather; and on section there is considerable resistance to the knife, showing more toughness and density than is possessed by the healthy organ.

On making a section through the organ, the cortex is found to be much reduced in thickness, both relatively and absolutely, being but one-fifth to one-sixth as thick as the medulla, instead of one-third, or one-fourth, as in the normal. The color of the cut surface, when fresh, is brick-red. Several cysts may be observed in the cortical portion of the cut surface; these are due in part to circumscribed dilatations of the renal tubules, brought about by contractions and obliteration of their lumen at different points by development about them of new, dense, contracting tissue. In other instances these cysts are the result of dilatation of Bowman's capsule through obstruction to the egress of the urine caused by obliteration of the renal tubules near the glomeruli. The Malpighian bodies are diminished in size, being scarcely discernible by the naked eye; and they are less numerous than in the normal organ. The cones are less affected, but in advanced cases some contraction is observable. White lines are often seen in the straight tubes of the medulla, which are deposits of urate of sodium, and sometimes this is also observed between the tubes. The straight tubes may contain colloid casts.

148 CIRRHOSIS OF THE KIDNEY.

Microscopic.—Examinations are best conducted by hardening pieces of the organ in alcohol, and making longitudinal sections which include the capsular edge. These may be stained in picro-carmine, or logwood. Under a low power (× 20 to × 50) it will be seen that the capsule

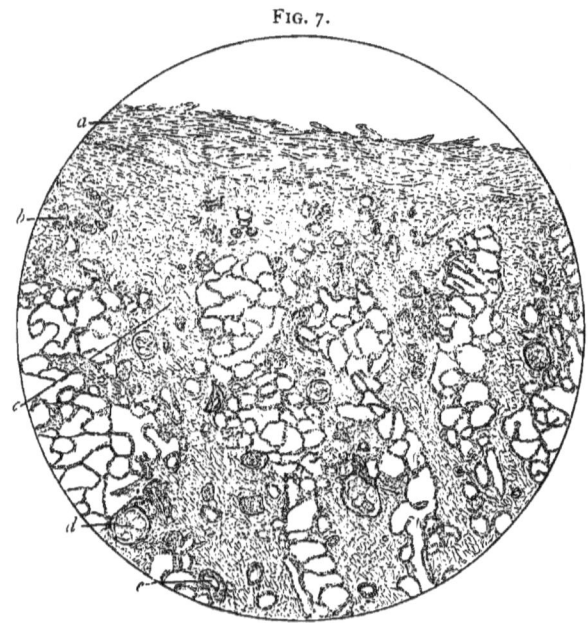

FIG. 7.

Cirrhosis of the kidney; longitudinal section of cortex, including capsule. *a*. Thickened capsule. *b*. Compressed tubule, nearly obliterated. *c*. Wedge-shaped band of dense tissue extending inward from capsule. *d*. Malpighian tuft somewhat compressed. *e*. Section of artery, showing thickened coats.

is thickened and adherent in places to triangular-shaped areas of the cortex, which point inward, the base of the triangle being presented to the capsule. These triangular patches of the cortex present a dense, homogeneous, granular appearance, in which very little of the outlines of the normal gland structure can be made out. The interlobular arteries may be made out, unusually tortuous in their distribution. On either side of the interlobular arteries may also be seen the Malpighian bodies; and the latter will be noticed to be very irregular in size, and situated nearer the trunk of the artery than normal. We observe, also, that the Malpighian bodies are occasionally situated immediately beneath the capsule of the organ, which never occurs in the normal. The medullary rays (bundles of convoluted tubes) partake of the same irregularity and tortuosity in their course, as the interlobular arteries already described; and this is due to the lessening of their field of distribution through contraction of the cortex. The triangular patches already described as pointing inward from the capsule are met by similarly shaped patches point to point, which pass out from the medullary border in the course of the interlobular arteries.

Under a high power ($\times$ 300 to $\times$ 400), it will be observed that the dense triangular patches before spoken of consist mostly of tubules and Malpighian bodies, in various stages of contraction and atrophy, between which is considerable increase of connective tissue. These cellular elements are chiefly embryonic, though in some places they are distinctly fibrillated and organized, especially in advanced atrophy of the organ.

The Malpighian bodies found in the section may present three varieties of appearance: (*a*) composed of a succession of fibrous laminæ which encircle the atrophied tuft; (*b*) the capsule of Bowman greatly thickened, but the capillary tuft only slightly altered; (*c*) the capsule of Bowman thickened and stretched, forming the so-called cyst; and the tuft pushed to one side, more or less atrophied, and filling only a part of the cavity.

The renal tubules are found in several conditions

according to their location, and the stage of the disease. Between the dense triangular patches little change may be discoverable. The epithelium of the tubules especially may present no perceptible variation from. the normal—indeed, it seldom does in the early stage of the disease. If, however, the disease is advanced, tubules may be seen which have shed their epithelial lining. Within the dense triangular patches are found more or less atrophied tubules, some of them shrunken to mere lines. The epithelium may contain fat globules, but desquamation does not occur to any extent. The tubules rather wither up, apparently from defective nutrition; or their walls may be seen thickened, and, like Bowman's capsule, blended with surrounding fibrous elements; or some of the tubes may be seen plugged with colloid matter. In the straight tubes of the pyramids, no epithelial changes are observable, but they may contain urate of sodium as before mentioned.

The walls of the arterial vessels are seen to be thickened. This thickening is not a simple hypertrophy of the muscular fibres; but it is rather an increased development of the connective tissue elements between the muscular fasciculi. The elastic fibrils of the intima or inner coat are also thickened.

The essence of the pathological changes found in the kidneys as the result of cirrhosis is increased growth of fibrous connective tissue between the convoluted tubes; thickening of Bowman's capsule; atrophy of the Malpighian bodies and convoluted tubes; and thickening of the muscular and fibrous coats of the small arteries. The ultimate result of these changes is contraction and atrophy of these organs.

Anatomical changes occur in other organs as the result of renal cirrhosis. The most constant of these are alterations in the tunics of the small arteries, similar to those described as occurring in the kidneys. The fibro-muscular change in the arterial tunics leads to degenerative processes, which weaken their structure and give rise to hemorrhages in various locations through giving way of the vessels; and this occurs most often in the brain. Hypertrophy of the left ventricle of the heart is a frequent accompaniment of this disease. It is present in the majority, and some claim in all well-marked cases, if the disease lasts a sufficient length of time.

An interesting post-mortem appearance is often observed in these cases, consisting of thickening of the dura mater, which is adherent to the vertex of the cranium; the latter also is thickened considerably, while the grooves for the temporal arteries are much deeper than in the normal condition.

Retinal changes are present in a large number of cases, probably in the majority at some time in the course. These will be noticed in detail under the head of symptomatology.

SYMPTOMS.

It is of the utmost importance to recognize renal cirrhosis in its earliest stages, for if not discovered until well established, we can hope for only temporary improvement from treatment. Renal cirrhosis in its early stages develops insidiously, being accompanied by but few pronounced symptoms. There is every reason to believe that the disease exists in some cases for years before it causes disturbances sufficiently marked to attract the attention of the patient. Indeed, it taxes the most acute skill of the physician to unmask this treacherous malady before the kidneys are gravely and irreparably damaged; and, therefore, the most systematic and careful scrutiny of the minute symptoms and signs which accompany the early course of the disease cannot be too earnestly insisted upon.

EARLY SYMPTOMS.—These are often vague and unpronounced, and vary much in different cases. Bearing in mind what has been said of the causes of the disease, we should first note the age and habits of the patient as to eating, drinking, etc. The family history should next be inquired into, and any occurrence of gout or apoplexy duly noted. By such means we often come upon some circumstances which indicate that the patient is predisposed to the disease. Suspicion having been aroused, we may look for the following symptoms.

Dyspepsia.—The patient loses his appetite, and is troubled with flatulent dyspepsia. In exceptional cases it may be absent, or it may be postponed till the disease becomes well developed. In a large number of cases, however, it is the first symptom to attract the patient's attention.

Neuralgia.—This is a pretty constant early symptom, most often occipital, though it may be vertical, or even frontal. In most cases neuralgia resulting from renal cirrhosis presents some peculiar features which should be remembered. It usually occurs in those who have been previously exempt from such disorders. The pain is persistent rather than severe, and usually extends down the back of the neck. It does not, like other forms of neuralgia, tend to manifest evening exacerbations; on the contrary, it is likely to be most severe in the morning, or during the early part of the day, and to subside toward evening. Diminution or *complete loss of sexual desire* occurs early in the disease. The exceptions I have met with to this rule have been rare, and confined chiefly to young patients.

Vertigo.—In some cases vertigo occurs quite early. It is usually of a transient, but quite pronounced character;

if the patient be standing or walking when attacked, he may stagger, or even fall. It may pass off in a few minutes, or continue for several days; but, having once occurred, it is liable to reappear repeatedly.

The Urine.—Very often the first symptom the patient observes is the fact that he rises at night once or oftener to urinate. This is due to the singular circumstance that usually more urine is secreted at night than during the day. This is especially the case later in the course of the disease; in fact, it is the rule to the end. In the early stages, the urine will be found to present the following characters. It is pale in color, of low specific gravity—1.014 to 1.018, is acid in reaction, and up to or above the normal in quantity. The urine may contain albumin, usually not more than a fraction of one per cent. The presence of albumin is exceedingly variable in this disease, especially so in its early stages. It may appear temporarily and then be absent for longer or shorter intervals of time. Its presence when suspected should be sought for repeatedly, and under circumstances especially favoring its excretion, as after exercise or eating.

When able to secure only a single sample of the urine for examination, I prefer that passed before retiring at night, as it is then more concentrated, and more likely to contain traces of albumin when the latter is transitory, and only present as the result of the disturbing influence of exercise, or the ingestion of food.

The urine may contain casts, and, if so, they are sparse, and of the small pale hyaline variety, possibly an occasional granular cast being observed.

Lastly, the urine in these cases usually throws down envelope-shaped crystals of calcium oxalate. Free uric acid crystals are scarcely less common, as the urine is loaded with urates in the early stage.

Circulatory System.—Among the most valuable, and certainly the most constant of the early symptoms of cirrhosis of the kidney are those which are referable to the vascular system; for on the blood seem to fall the earliest consequences of defective renal function. In searching then for the early indications of this disease, the circulatory apparatus should receive close scrutiny, as evidences of the disease are always present there, even in the inceptive stages.

The pulse is characteristic; it will be found hard and resistant, "rolling under the finger like a cord." It is prominent, and easily found on passing the fingers across the wrist; on compressing the artery firmly it will be found more difficult to obliterate the pulsations than normal, in consequence of the increased vascular pressure. These characters of the pulse are always present, though in varying degrees, and although we cannot accurately estimate the characters of high pressure by the finger, we can do so by means of the sphygmograph. Since then this instrument affords such valuable information in these cases, I will briefly describe the characters of sphygmograms which indicate high arterial pressure. To

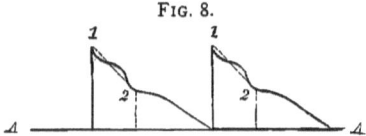

FIG. 8.

the late Dr. Mahomed, of Guy's Hospital, we are indebted for many valuable suggestions on this point. His method of gauging high arterial pressure is as follows: A line must be drawn from the apex of the upstroke of the tracing (1, Fig. 8) to the bottom of the notch (2) preceding the dicrotic wave. If any part of the tracing rise above this line (1, 2), then the pulse is one of high pressure. The height of the notch (2) may also be taken as an indication of the pressure; the higher it is from the respiratory line A A, the higher is the pressure.

On comparison of normal tracings (Figs. 9 and 10) with those of cirrhosis of the kidney (Figs. 11, 12, and 13), the

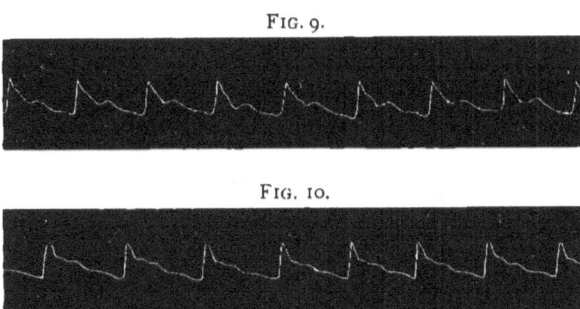

Figs. 9 and 10. Sphygmograms of normal pulse.

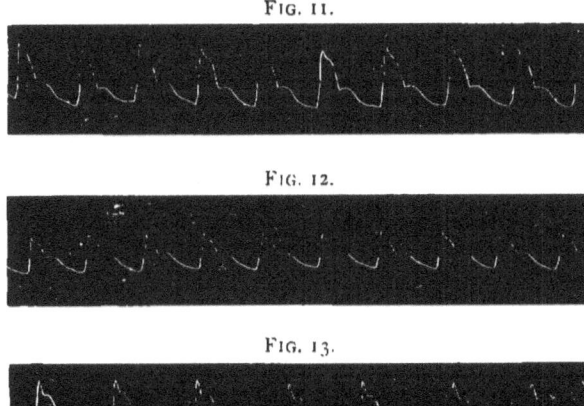

Figs. 11, 12, and 13. Sphygmograms taken from cases of cirrhosis of kidney.

differences will at once appear as striking. Now the characters of high tension in the pulse just described and illustrated, as shown by the sphygmograph, I have never found to be absent in the early stages of cirrhosis of the kidney. It is true we may meet with high arterial tension in a few other conditions, as in atheromatous states of the vessels, and in nervous states which cause vaso-constriction of the small vessels (arterioles), as in hysteria, etc.; also, as I have recently shown (*Journal of American Medical Association*, September 12, 1885), high tension usually accompanies diabetes insipidus and mellitus; but it is usually easy to exclude most, if not all of these, and therefore this symptom may be justly looked upon as one of great value.

The Heart.—More or less hypertrophy of the left ventricle of the heart is usually present in renal cirrhosis. This may not occur till the disease has been in progress some time, and hence it is not intended to include this among the earliest indications of the disease, although it may be so. As the causes which bring about hypertrophy of the heart are in operation from the beginning, we shall be able to find evidences of its increased power by other means than the sphygmograph. Thus the heart sounds are usually abnormally loud; the second or diastolic sound is especially accentuated, as heard to the right of the sternum in the interspace between the second and third costo-sternal cartilages.

Epistaxis.—This is another evidence of the increased power of the heart's action that often occurs quite early in the disease. It is characterized by great obstinacy, sometimes persisting for several days. In my experience, it is an exceedingly frequent symptom. I have known it to be present in several cases *two* and *even three years* before pronounced symptoms of the disease could be made out. We may inquire for this symptom especially in those cases in which a tendency to plethora exists; in the opposite condition of the system this symptom is usually absent.

Such, in the main, are the symptoms to be sought for in the early stage of the disease. We shall seldom find them all present, but usually a sufficient number may be made out to arrive at a correct conclusion. While the urine should receive the closest attention, too much reliance must not be placed upon the absence of albumin therefrom; for it should be borne in mind that cirrhosis of the kidney has its *non-albuminuric stages*, even when the disease is well developed, but these are especially likely to be present in the early course of the disease. I hold, indeed, that cirrhosis of the kidney may be recognized without taking into consideration the presence or absence of albumin; although it may not *always* be possible to do so, owing to variability of the symptoms. But if we have, for instance, evidences present of high arterial pressure (excluding other causes); an increased quantity of pale urine, of continued low specific gravity; the patient rising at night to urinate (as a recent habit); occipital headaches; dyspepsia of a flatulent character and recent origin, without obvious cause or previous attacks; and hypertrophy of the left heart, in a person of middle age, there can no longer be doubt that we have to do with a case of renal cirrhosis.

MATURE SYMPTOMS.—When the disease becomes fully developed, most of the symptoms already described become more marked; and, moreover, several additional ones sooner or later arise. It will be convenient to deal with these systematically.

The Digestive System.—The dyspepsia spoken of in the early stages now becomes aggravated, and is of a notably flatulent character; nausea and vomiting may also be superadded, and are most common in the morning on rising. Diarrhœa is sometimes associated with the above symptoms. These gastro-intestinal disturbances are not constant; they may occur and continue for a few days or weeks, and then pass off, or in some cases they do not appear till very late in the disease; exceptionally they are altogether absent. In the very late stages of the disease, another form of diarrhœa is present which will be considered at the proper time.

The Urinary System.—The urine now becomes noticeably increased; it is considerably above the normal in daily quantity; its specific gravity sinks still lower—1.010 to 1.014. The urea and phosphates suffer the greatest reduction. The absolute reduction in the quantity of urea is characteristic throughout the course of the disease, though this varies at different periods considerably, becoming more pronounced with the advance of the disease. The reduction is often marked just before the outbreak of acute uræmic symptoms. Under such circumstances the quantity of urine, also, becomes reduced; it may fall considerably below normal, thus causing a rise in the specific gravity, though the latter seldom if ever reaches the normal standard. The color of the urine becomes lighter—as pale as water, and its characteristic odor is faint or imperceptible. The urine is clear and usually deposits but little sediment. The quantity of albumin when present varies from 0.2 to 0.5 per cent.; it may be temporarily absent as in the early stage, but as a rule it will be found if sought for over a sufficient length of time. The sediment contains scattered casts. In addition to hyaline casts, granular ones are usually now to be found, as well as those containing a few isolated fat globules; but both of these are much less frequently met with than the small hyaline casts. The sediment also contains more or less leucocytes and epithelium in various states of maceration. Lastly, oxalate of lime and uric acid crystals in varying numbers are usually to be seen, as in the earlier stages of the disease.

The Respiratory System.—Dyspnœa is perhaps the most common symptom of the disease referable to the respiratory system. This may occur under three forms, the most common of which at this stage of the disease is that resembling asthmatic attacks in many respects. Thus, it is likely to appear suddenly, most often at night, and the attack may continue for several hours, during which the patient experiences great distress in breathing; the respirations are accompanied by sibilant râles, and, finally, the attack may terminate with more or less watery expectoration, just as in typical asthma. It is probable that this is a nervous manifestation of uræmic origin. The second form of dyspnœa ushers in œdema of the lungs, and this is a very fatal complication of the disease. In this form there is no interval, or remission; once the embarrassment to respiration sets in, it gradually increases, the respirations become quickened—forty to sixty in the minute—and, in short, all the evidences of *œdema pulmonis* become developed. While this can scarcely be considered among the early symptoms of the disease, yet in not a few cases it is the first warning the patient experiences. The origin of this symptom being uræmic, it will be readily understood that it may arise at almost any period in the progress of the disease. The third form of dyspnœa met with is the result of dilated and weakened conditions of the heart; and hence it only arises in very advanced stages. The characteristic of this form is, that it is chiefly excited by exercise ; the patient experiences but little, if any, distress if he remain quiescent. In short, it strictly conforms to the symptoms common to dilated and fatty heart, upon which, in truth, it depends.

Other disorders of the respiratory system occur in cirrhosis of the kidney—in fact, the whole respiratory tract is very prone to irritation of an inflammatory tendency in the course of this disease. Bronchial, pharyngeal, and naso-pharyngeal catarrh are frequent accompaniments of cirrhotic kidneys ; and we sometimes meet with these early in the disease. Cheyne-Stokes respiration is common, more especially in chronic uræmic conditions.

Lastly, œdema of the lungs and pneumonia are frequent and often fatal complications. This is because they are excited by the irritating qualities of the blood, which is loaded with elements of tissue waste, and retained renal excreta ; and, moreover, once inflammatory changes are set up, they are fanned by increased tension in the arterial circulation.

The Vascular System.—We have already spoken of the characters of the pulse in this disease, and of the increased arterial pressure that so constantly prevails, so that nothing remains to be added concerning that symptom here. A most remarkable chain of cardiac complications arises in the great majority of these cases, which merits special consideration. The first step in these changes consists of hypertrophy of the left ventricle of the heart, already referred to. This becomes more marked as the disease advances. The area of cardiac dulness becomes more and more increased; the apex beat becomes displaced, and appears in the region of the seventh, eighth, or ninth left rib, beneath the mammary gland. The impulse of the left ventricle becomes more and more powerful, the chest walls visibly heaving with each contraction. The sounds of the heart, especially the second, become louder than normal; and they may be reduplicated. As the disease becomes far advanced, however, these conditions gradually undergo change, and at length a stage is reached when the heart is unable to meet the increasing demands made upon it, and dilatation of its hypertrophied ventricle commences. As the latter advances, the pulse loses its fulness, and the tension in the arteries is lessened. The urine now diminishes in quantity, and other symptoms set in, which will be described later. The last step in these cardiac changes is a degenerative one: if the patient survive long enough, fatty changes in the muscular walls of the heart become established. Carditis—a rapidly fatal complication—is one of the secondary inflammations common in the course of the disease.

It will be readily understood that changes of such a marked character, occurring in the central organ of the circulation, must result in disturbances of the vascular system elsewhere. Some of these have already been mentioned, as increased arterial tension, epistaxis, etc. It remains to speak of others, some of which are far more serious; the most prominent is apoplexy. During the stage of hypertrophy of the heart, before dilatation sets in, it is not at all uncommon for these patients to be stricken down suddenly with cerebral apoplexy. As was seen in the consideration of the pathology of this disease, the small arteries undergo atheromatous and degenerative changes, which greatly weaken their walls; and the increased power of the heart renders them very liable to rupture, and consequently hemorrhages are a common feature of the disease. The cerebral vessels not only possess more delicate tunics, but receive less support from surrounding tissues than most others; and consequently they are the most likely to give way in these cases. The vast majority of cases of cerebral apoplexy owe their origin to cirrhosis of the kidneys. For the same reasons, hemorrhages into the retina are frequent, resulting in more or less complete amaurosis. These will be considered when the retinal changes are dealt with in detail. Less frequently, hemorrhages occur in the stomach or other viscera.

As signs of the disease referable to the vascular system, may be mentioned great tortuosity of the superficial arteries, especially the temporal. I have likewise frequently noted in these cases small circumscribed aneurisms in such vessels.

Whether the hypertrophy of the heart in this disease, is due to the resistance in the capillaries or tissues, caused by the irritating qualities of impure blood; or whether the cause operates directly upon the heart itself, exciting it to increased power and development, are questions which are still undecided; and consequently their discussion, however interesting, does not concern us in a practical treatise such as this is intended to be. It may be said, however, that the evidence favors the view that the renal defect is the primary element in the cause, to which all others are secondary.

The Muscular System.—The consequences of renal cirrhosis do not always fall upon the muscular system to an appreciable extent till the disease has been in progress a considerable time. In exceptional cases there may be quite a noticeable diminution of muscular power in the early stage of the disease. This is more likely to be the case with elderly people, or those whose general condition was previously impaired. As the disease becomes developed, the muscular system presents evidence of impairment. While most patients maintain their strength apparently little impaired until quite late, in a few cases debility gradually creeps over them, and they become more and more conscious that they are invalids. Finally, the muscular system suffers from imperfect metabolism; its nutrition becomes impaired from alterations in the blood which weaken its nourishing qualities; and the patient may lose in weight, though emaciation is rarely extreme.

The Cutaneous System.—In this, as in most other renal diseases, the skin feels dry and harsh to the touch. It does not, as in acute forms of nephritis, become blanched and anæmic; though sooner or later it loses its healthy color and assumes a dirty pallor; added to this, *in the late stages* of the disease, is a yellowish, sub-jaundiced hue, the result of impaired hepatic function—more or less cirrhosis of the liver being present. Skin eruptions are not uncommon, more especially in the form of *acne rosacea*. Itching of the skin may be present, sometimes of a most inveterate character; and with this is often associated an erythematous rash.

The Nervous System.—Probably the most numerous class of symptoms resulting from this disease is that which becomes manifested through the nervous system. Some of these nervous phenomena are of trivial importance, while others are fraught with consequences of the utmost gravity. Nearly all of them are expressions of some form of uræmia. Among the less serious manifestations may be mentioned: disorders of hearing in some cases amounting to temporary deafness; insomnia, marked in some cases, and often appearing among the early symptoms; mental depression, and hypochondria; neuralgic pains, which, in addition to those so common in the head, occur in various locations, but especially in the lower extremities; numbness and temporarily impaired sensibility in certain nerves, sometimes quite extensive, and of the hemiplegic order. But the most serious nervous derangements are coma and convulsions, which are met with more or less frequently throughout the whole course of the disease; indeed, they are among the most commonly fatal terminations thereof. A detailed description of these convulsive and comatose disturbances is given in the chapter on uræmia in the early part of this work.

Disorders of vision are exceedingly common in renal cirrhosis. They may be but slight, or they may amount to complete blindness; they may appear suddenly within a few hours, or they may come on gradually, their development extending over weeks or months; they may be quite transitory, passing off in the course of a few hours or days, or they may be of a permanent and incurable nature. The slighter forms of these disturbances are such as to attract but little attention: consisting of apparent specks floating before the eyes, or the appearance of minute bright sparks disappearing so rapidly that the vision fails to fix distinctly their impressions.

Uræmic Amaurosis.—The more pronounced impairment of vision occurs under two forms: uræmic amaurosis and "albuminuric retinitis." Uræmic amaurosis, as its name implies, is of uræmic origin; it is of sudden onset, and consists in complete loss of vision, although the reaction of the pupils is preserved. This disturbance quickly disappears with improvement of the accompanying uræmic phenomena; sometimes in a few hours or days, and no ophthalmoscopic changes are discoverable in the retina, save those which may have previously existed.

"*Albuminuric Retinitis.*"—From the frequency of occurrence, as well as the marked degree of visual impairment, this form merits more minute consideration.

Eales's statistics (*Birmingham Medical Review*, January, 1880) of one hundred cases of chronic renal disease, show that retinal changes occurred in 28, or 1 in 3½ of the cases. It not infrequently happens that the impairment of vision first leads the patient to seek medical advice, so early may this disorder arise. The character of the symptoms varies widely in different cases, owing to the differences in the nature of the retinal changes present. Sometimes the impairment of vision develops gradually, and may not become marked for several months; while in other cases complete loss of sight takes place suddenly. All of these changes are susceptible of improvement up to a certain point; but complete restoration of vision is seldom attained. The visual function, therefore, exhibits a varying range of impairment: in some cases near vision remains comparatively good, the patient being able to read, or even write well; while objects at a distance appear indistinct, blurred, or distorted.

In other cases near objects appear indistinct, or the vision easily tires, and fails to fix the object viewed long enough fully to comprehend it. Thus, in attempting to write, as one of my patients expressed himself, "the last letters of the word become blurred, and I find, on finishing, that they are scrawled out irregularly over the paper." When these disorders of vision become established—from retinal changes—they are aggravated on the approach of acute uræmia; and so repeatedly will this be found the order of events, that I am now in the habit of carefully comparing the capacity of vision for evidences of approaching uræmia.

The ophthalmoscopic changes in the retina are lucidly described by Dr. Gowers in his *Medical Ophthalmoscopy* (second edition, London, 1882), with whose kind permission the following brief synopsis is appended.

The retinal disease presents certain elements which are variously combined in different cases. These are: (1) diffuse slight opacity and swelling of the retina, due to œdema of its substance; (2) white spots and patches of various size and distribution due, for the most part, to degenerative processes; (3) hemorrhages; (4) inflammation of the intra-ocular end of the optic nerve; (5) the subsidence of inflammatory changes may be attended with signs of atrophy of the retina and nerve.

In most cases one or the other of these changes predominates, especially in the early stages of the affection, and, according to the element most conspicuous, four types of disease may be distinguished. These are—the degenerative, the hemorrhagic, the inflammatory, and the neuritic—according as white spots of degeneration, extravasations of blood, parenchymatous retinal inflammation, or inflammation limited to the optic nerve, predominate.

The *degenerative form* is the most common. It commences usually without signs of inflammation, by the appearance of small white spots on the substance of the retina, sometimes near the optic nerve entrance, sometimes at a distance. They are commonly at first soft-edged and rounded, and as they enlarge become irregular. Generally, small white spots, often punctiform or elongated, are seen around the macula lutea, arranged in a radiating manner, although frequently not forming a complete circle. Often a less intense and diffuse opacity is visible in tracts here and there. Hemorrhages, constant in all varieties, are slightest in the chronic degenerative forms. They often are adjacent to the white spots due to changes in the nerve fibres, and, lying for the most part in the nerve fibre layer, they have a more or less striated arrangement, determined by the nerve fibres, the direction of which the striæ follow. Sometimes linear hemorrhages are seen. When large, the extravasations are flame-shaped. The retinal changes in this form may be considerable without any alteration in the optic disk; often, however, its edges are blurred, the physiological cup indistinct, and the tint abnormal—reddish-gray.

In the *hemorrhagic form* the conspicuous change is the occurrence of a large number of hemorrhages, with but little degenerative change and but slight signs of inflammation of disk or retina. Commonly, after a time, there is more or less degeneration adjacent to the hemorrhages, and traces of the halo of spots around the macula are rarely absent. The hemorrhages, for the most part, resemble those just described, differing only in number and size.

In the *inflammatory form* there is a general parenchymatous swelling of the retina with complete obscuration of the disk. The vessels are concealed, the arteries especially. The veins are distended, and sometimes have an extremely irregular and tortuous course over the fundus; the *arteries are narrow*. Hemorrhages occur in considerable number, and are often large and striated. White spots are commonly numerous, and more or less uniform in character, especially in acute cases, when they are large, rounded, and soft-edged.

In the *neuritic form* the inflammation of the optic nerve predominates over the other retinal changes to such an extent that it may appear to be the sole alteration, and may present nearly the aspect which is common in intracranial disease. The edges of the disk are veiled under a grayish-red swelling of moderate prominence, which may extend a little distance beyond the normal edges of the disk. The prominence may be slight, or such that the veins form conspicuous curves over the sides. The arteries are usually narrow, and often concealed in the swelling; or even the veins may be concealed. A careful examination will show, in almost all cases, signs of slight retinal degeneration, sometimes so slight as to require close attention, and careful focussing by the direct method to detect them. It is remarkable that there is little tendency for hemorrhages to occur in the swollen papilla in this form. Before quitting this subject I desire to draw especial attention to the appearance of the retinal vessels frequently present in renal cirrhosis, as first pointed out by Dr. Gowers. In such cases the veins are not larger than normal, but the arteries *are not more than one-half or even one-third the diameter of the veins*, instead of being two-thirds, or three-quarters the diameter. "The comparison is to be made between arteries and veins which run side by side, and correspond in distribution." This condition of the retinal vessels, observed in cirrhosis of the kidney, often occurs very early in the disease, and in some cases independently of any other retinal change; hence it is important as an early diagnostic symptom.

LATE SYMPTOMS.—It remains to notice certain symptoms characteristic of the advanced stages of the disease. The complications likely to arise during the progress of the disease are so numerous that the majority of patients succumb to some of them before this stage is reached; indeed, the physician may congratulate himself if his skill has enabled him thus far to prolong the patient's life.

Dilatation and degeneration of the cardiac structure have now set in, and the abnormal force of the heart's action has passed away. The pulse has lost its tension, and become weak, small, rapid, and perhaps irregular and fluttering. In consequence of the reduced arterial pressure, the tension in the renal vessels falls, and the urine becomes diminished in quantity. The solids also of the urine remain diminished, and the specific gravity continues below normal; though as the quantity of the urine suffers marked reduction the specific gravity may mark some increase, rarely, however, reaching the normal standard. The consequences are that uræmic symptoms almost constantly threaten the life of the patient.

As a result of the failure of the circulation, dropsy sets in; beginning usually in the feet and extending upward, it may involve the abdominal cavity; though if there be cardiac improvement its progress may be arrested.

Still another consequence of the impairment of the circulation is the development of cirrhosis of the liver, evidenced by a sub-jaundiced hue of the skin. This, in turn, obstructing the portal circulation, often causes hemorrhoids. Obstinate diarrhœa may harass the patient, accompanied by ulceration of the intestinal mucous membrane; but more often an œdematous condition is present, and the stools are watery and often streaked with blood. A singular circumstance in connection with the diarrhœa in cases of late cirrhosis of the kidney is that the stools are passed most often at night—as is also the greater part of the urine. I have, indeed, known a patient to pass a whole day without any, or at most one or two movements of the bowels; while at night from six to ten watery stools would be passed; and this order of events continue for several months.

In this stage also the degenerative changes in the retina are most common, resulting in gradual impairment of vision.

Uræmic symptoms, in the form of headache, pains in the limbs, or temporary numbness in certain nerves, are almost constantly present. The patient is unable to indulge in exercise to any extent without becoming "out of breath." There is loss of flesh—more apparent if dropsy be absent, or temporarily relieved by treatment.

At last a period in the history of the disease arrives when the contraction of the kidneys becomes so extreme that sufficient renal structure no longer remains unimpaired to maintain the urinary function; the urine falls much below the normal in quantity; diuretics fail to relieve the oliguresis, and coma or convulsions set in, most often the former; or, some of the secondary inflammations within the chest are kindled, which quickly result in death.

COURSE AND DURATION.—Cirrhosis of the kidney is essentially a chronic disease from its beginning. Indeed, it steals on so slowly, that it has been asserted that in some cases half a lifetime is occupied in leading up to its full development. While this statement is probably an exaggerated one, it is not at all uncommon for the disease to last from three to five years before it is recognized. If the disease run its natural course, without death overtaking the patient as the result of some of its many complications, it may not prove fatal for ten or fifteen years; and a few cases are on record in which it extended twenty years. The duration of the disease will depend largely upon the presence and nature of its complications; but scarcely less so upon its management.

In the early course of the disease the patient is seldom disqualified for his usual daily pursuits, and the general health may not seem deteriorated, either to the patient himself, or to his friends. This condition of matters may continue for months, in some cases for years, without any positive manifestations of the disease. It is for these reasons that the disease frequently escapes detection in its early course.

Sooner or later, however, some of the more noticeable manifestations of the disease attract the patient's attention. He rises more or less frequently at night to urinate; he loses his appetite; and dyspeptic symptoms, perhaps for the first time in his history, arise; or it may be that some impairment of vision caused by retinal changes attracts his attention and leads him to seek medical advice, usually without suspecting its cause.

Up to the period of the development of hypertrophy of the left heart, the symptoms chiefly consist of uræmic disturbances, such as headache, dyspepsia, visual disorders, etc.; and these are the more likely to arise at this time, since the absence of hypertrophy of the heart renders that organ unable to overcome the defect in the renal function. The urine is less uniformly maintained above normal, and is also liable to become diminished in consequence of some accession to the disease, or some systemic "storm" which develops the latent renal defect. In the absence of these, the manifestations of the disease are so slight, that it is often difficult to convince the patient that anything serious "is the matter with him." The urine may contain albumin or it may not, as albuminuria is an inconstant symptom in the early history of this disease. The urine may be above normal in quantity, and lighter in color, though not to so marked a degree as during the stage of cardiac enlargement. It is quite common during this time for the morbific changes in the glands to come to a temporary halt in their progress. If this be taken advantage of by judicious medication and proper hygienic measures, the disease may remain quiescent for months or even years.

After the cardiac enlargement becomes fully established, as a rule, the disease pursues a more rapid course; though even in this case, life may often be greatly prolonged by judicious measures. Danger now threatens the patient from the abnormal intravascular pressure in the arterial system. Hemorrhages may take place into the retina, brain, or other organs. The patient may still be out and about daily, and to his friends he may seem little changed. A slight exposure may result in the contraction of cold, and at first apparently trivial bronchial trouble may be hurried on by the great force of the arterial circulation, and the irritating waste products in the blood, into a rapidly fatal pneumonia. Or it may be that exposure, or some imprudent indulgence in wine, causes an acute extension of the disease in the kidneys themselves, which so cripples their functional capacity that acute uræmia is precipitated as a result, and this may quickly terminate life. It may be, on the other hand, that the disturbing influences just spoken of add suddenly irritating qualities to the blood, and secondary complications are kindled—carditis, pleurisy, œdema of the lungs, peritonitis, etc.; any or all of these are liable to prove quickly fatal. In the absence of secondary complications, however, the muscular structure of the heart at last weakens under the continued stress to which it has been subjected, and dilatation, usually with fatty degeneration, ensues. The patient now becomes invalided, and is unable to go about without effort and fatigue. The urine falls below normal in quantity, and the patient's life becomes a continuous struggle against uræmia in one form or another. Diarrhœa of an obstinate character often sets in, still further reducing the strength; the pulse grows weak and fluttering; the respirations are quickened on slight exertion; and at last the patient succumbs to some of the consequences of extreme atrophy of the organs.

DIAGNOSIS.

CONSTRUCTIVE.—This is especially important in the early stage of the disease, as probably the majority of these cases are overlooked until the disease is well established. It is only recognized in its earliest development through the utmost watchfulness, combined with a thorough knowledge of all its accompanying phenomena. I would lay especial stress on the necessity of habitually endeavoring to trace every symptom to its source in cases which present the slightest evidences of the disease. We should familiarize ourselves with the class of people in whom the disease is most likely to arise. The age of the patient, if over forty, should first be noted. Next, what kind of life has he led? What have been his habits? Has he lived freely, habitually overloading his stomach with meat and nitrogenous foods, and drinking much spirits? Has he suffered from gout; or has this disease or apoplexy constituted a feature of his family history? If such points are carefully considered, together with the early symptoms already described, the disease will not long elude discovery, even in its earliest stages.

DIFFERENTIAL.—The diseases of the kidneys most likely to be confounded with cirrhosis are, chronic nephritis and lardaceous disease.

The chief distinguishing points between cirrhosis of the kidney and chronic nephritis are as follows: In chronic nephritis, the presence, or a recent history of dropsy, is usually ascertainable; the urine is not increased in quantity—it may be diminished; albumin is usually present in considerable quantities; the urinary sediment is considerable; broad hyaline, granular, and fatty casts are numerous; anæmia is always marked; hypertrophy of the heart is rare, as are also evidences of high arterial pressure. Such are the chief characteristics of typical chronic nephritis, which differ essentially from those of renal cirrhosis.

In the very late stages of chronic nephritis contraction of the kidneys may occur, accompanied by hypertrophy of the heart, when it will be much more difficult to distinguish between the two, as the symptoms in both will then be very similar. In such cases we must rely largely on the history, which differs essentially. Thus, in chronic nephritis there is often a history of a preceding acute attack; dropsy will have been present at some period in its course, with profuse albuminuria; the hypertrophy of the heart rarely attains such prominence in chronic nephritis, as in true primary cirrhosis; the urine is less markedly increased in quantity in the former; and lastly, anæmia is always a more prominent feature in chronic nephritis than in renal cirrhosis.

The points of distinction between lardaceous disease and renal cirrhosis are chiefly as follows: In the former dropsy is frequently present; the urine contains a relatively large amount of albumin; the casts are broad, hyaline, waxy, and sometimes react with iodine characteristically; hypertrophy of the heart is rare, as is also high arterial pressure; obstinate diarrhœa is present from a comparatively early stage; uræmic disorders are rare; marked cachexia is present; enlargement of the liver and spleen is usually present; and lastly, a frequent history of syphilis, phthisis, or some chronic suppurative process, serve to form a combination of characteristics in typical cases of lardaceous disease which is essentially different from that of primary cirrhosis of the kidney.

PROGNOSIS.

DIFFERENTIAL DIAGNOSIS.

	CHRONIC NEPHRITIS.	CIRRHOSIS.	LARDACEOUS DISEASE.
Age	Occurs mostly before 40.	Occurs mostly after 40.	Most frequent in childhood and early life.
History	Frequently preceded by an acute attack.	Gout. Lead. Apoplexy. Generous living.	Syphilis. Phthisis. Scrofula. Chronic suppurative processes.
Distinguishing symptoms	Dropsy general. Features puffy and pallid. Anæmia prominent. Hypertrophy of heart rare. Uræmic disturbances rare, save very late. Amaurosis infrequent.	Dropsy absent, save in last stages. General condition well preserved. Hypertrophy of left heart commonly present, also high arterial pressure. Amaurosis common. Uræmia frequent in all stages.	Dropsy common. Diarrhœa persistent. Uræmia rare, save at close. Cachexia prominent. Enlargement of liver and spleen usual. Heart rarely hypertrophied. Amaurosis uncommon.
The urine	Not increased in quantity. Albumin large in quantity. Epithelial, fatty, and granular casts. Phosphates slightly reduced. Chlorides diminished considerably. Urea slightly diminished. Indican not increased.	Considerably increased in quantity. Albumin small in quantity. Small hyaline casts; occasionally granular. Phosphates greatly reduced. Chlorides reduced variably. Urea considerably diminished. Indican increased.	Increased in quantity. Albumin large in quantity. Broad hyaline casts; may react with iodine. Phosphates much reduced. Chlorides greatly diminished. Urea not much reduced till late. Indican sometimes increased.

PROGNOSIS.

The prognosis of renal cirrhosis will depend largely upon the period in which it is discovered, so far as the immediate outlook is concerned. Thus, if it be recognized in its early stages before hypertrophy of the heart has set in, it may be checked, and even brought to a standstill; at least for more or less extended periods of time. It is important to bear in mind that the disease has its periods of dormancy in its early history, and if these are taken advantage of and the patient placed under judicious hygienic surroundings, carefully guarding him against all those influences which are prone to excite acute augmentations of the disease, his life may be prolonged for years.

176   CIRRHOSIS OF THE KIDNEY.

When, however, hypertrophy of the heart becomes established, the outlook is more grave. The changes both in the kidneys and in the circulatory organs are then so far advanced that, as a rule, treatment fails to arrest them; and, sooner or later, the disease proves fatal. As unfavorable indications, we may note uræmic symptoms, especially convulsions; weakening of the heart's action, with irregular, fluttering pulse; short respirations, diarrhœa, etc. Yet we are scarcely ever justified by the severity of symptoms in predicting immediate dissolution; and certainly never in abandoning hope of improvement for a longer or shorter time. Indeed, there are few other diseases capable of presenting so formidable and grave symptoms, and at the same time being susceptible of improvement to the extent that the patient may get up and go about for months, or even years. I have at present a case in charge that well illustrates the above facts. Nearly *two years ago* he was taken with uræmic convulsions; previous to which he had been gradually

FIG. 14.

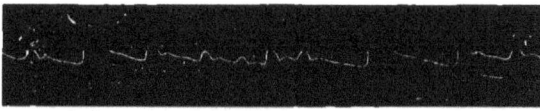

FIG. 15.

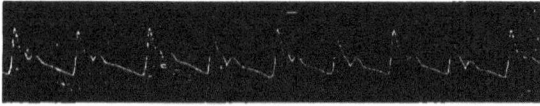

losing his sight. Following the convulsions œdema of the lungs set in; but both were brought under control. The heart was dilated, and the pulse so very irregular, that the sphygmogram taken was an irregular scrawl, as seen in Fig. 14. The cardiac muscle was strengthened by a more liberal diet and the persistent use of tonics, including iron, strychnine, digitalis, etc., until the sphygmograph indicated comparative regularity of the heart, as shown in Fig. 15. The patient is now so greatly improved that he is able to be about, and rides out daily in fine weather.

TREATMENT.

The general principles to be followed in the treatment of cirrhosis of the kidneys consist: (*a*) in modifying those conditions, both general and special, which seem to give rise to, or which tend to aggravate the disease; (*b*) in protecting the system from the consequences of defective renal function; (*c*) in limiting the damage already caused by the disease, both in the kidneys and in the organism; and so far as is possible in restoring the latter to its normal condition. Under the first head may be ranked the observance of the various hygienic measures detailed in the chapter on albuminuria. The second consists chiefly in the employment of measures calculated to relieve the various uræmic manifestations, as well as the dangers consequent on abnormally high tension in the arterial system. The third division consists largely in the employment of medicinal agents proper, which are directed against the disease itself.

As the type of this disease differs essentially in its various stages, the treatment, therefore, will vary accordingly; and, with the view to render the latter more intelligible, the treatment appropriate to each stage will be considered separately.

EARLY STAGE.—The special hygienic measures to be insisted upon, are: to limit the amount of nitrogenous foods to be indulged in, thus putting the organs functionally at rest, so far as is possible. This is especially important in this stage, and the enforcement of the rule should be rigid, unless the general condition of the patient urgently demands supporting measures, which is rarely the case. The clothing should be such as to prevent danger of chills to the skin, which are so liable to provoke exacerbations of the disease. The urine should be regularly measured by the patient at least twice a week, and if it should fall below the normal quantity, it must be brought up to a healthy standard; for which· purpose diuretic agents should be employed, such as five to ten drops of the tincture of digitalis, combined with one of the salts of potassium, sodium, or lithium—say the citrate of potassium, in doses of ten to twenty grains, three or four times daily.

It will be found that the vital functions will proceed with the least disturbance if the urine be maintained rather *above* normal in quantity. Diuresis usually takes place as a result of the disease; yet the urine may fall off temporarily from various causes; and if the oliguria continue, we are likely to have grave disturbances, which should be guarded against as just indicated.

The bowels must be carefully regulated, and if the pulse shows evidences of marked tension, they should on no account be permitted to become constipated. The systematic regulation of the bowels tends, probably more than any other measure, toward postponing the development of cardiac enlargement.

We should next resort to the use of those agents, the action of which is calculated to check the development of hyperplasia of the connective tissue in the kidneys. For this purpose the most popular remedies are the iodide of potassium, mercury, and the double chloride of gold and sodium.

The iodide of potassium is highly spoken of by both Bartels and Ralfe in these cases ; and, on the whole, it may be looked upon as a promising remedy, if given in the early stages, and continued over extended periods of time. It is best administered in medium doses—five to ten grains, three times a day, well diluted; and if digestion be weak, it should be given after food. Mercury is recommended usually in the form of bichloride, in minute doses (one-thirtieth of a grain), either in form of pill or solution, three times daily, after food. I have found the protoiodide of mercury a much more effective preparation, given in one-sixth of a grain doses after meals, in the form of pill. This preparation combines, to some extent, the virtues of iodine with those of mercury, rendering the latter more active; at the same time it is much less likely to induce ptyalism than the bichloride—a very important consideration in these cases, as the condition of the kidneys renders these patients peculiarly sensitive to the action of mercurials. The chloride of gold and sodium was first recommended in this disease by Dr. Bartholow (*Materia Medica and Therapeutics*), to be given in pill-form, in doses of one-thirtieth to one-twentieth of a grain, three times a day. Dr. Tyson gives it in larger doses, one-sixteenth to one-tenth of a grain. I have seen beneficial results follow its administration : the character of the urine improved ; nocturnal urination diminished—in one instance ceased; and other evidences of improvement became manifest. In my hands the chloride of gold and sodium has usually produced some unpleasant effects in doses exceeding one-twentieth of a grain, such as slight fever, uneasiness in the gastric region, etc. In doses ranging from one-thirtieth to one-twentieth of a grain, however, it usually improves the appetite and digestion, and regulates the quantity of urine passed relatively—diminishing the nocturnal secretion, and increasing that of the daytime.

It should be borne in mind that substantial benefit can be expected from the class of agents just considered only through their *long continued* use; their administration should therefore be continued for months. On the whole, I believe that considerable good may be expected from the use of these medicines now that we are enabled to diagnosticate the disease in its early stages, for it is in this period only that we can hope for favorable results.

STAGE OF CARDIAC ENLARGEMENT.—The indications for treatment in this stage of the disease are essentially different from those in the preceding. In addition to the dangers to be apprehended from impairment of the renal function, we now have those from great increase of pressure in the arterial system, consequent on enlargement and increase of power in the cardiac muscle. The pathological changes in the coats of the smaller arteries render them weak and liable to rupture, and consequently hemorrhages are not uncommon in this stage.

With a view to anticipate these accidents, we should enforce rest and low diet; and administer cardiac sedatives and those agents which lower the abnormal tension in the vessels.

These patients should not be permitted to indulge in much exercise, and they should especially be cautioned against running, or ascending flights of stairs, and 'such efforts as call suddenly on the power of the heart. The diet should be unstimulating; and special care should be exercised not to overload the stomach. The bowels should receive scrupulous attention, and be maintained in a soluble condition. If evidences of high arterial tension are very marked, as shown by great hardness with fulness of the pulse (better indicated by the sphygmograph), a sense of fulness in the head with vertigo, etc., a course of purgatives should be given until the abnormal tension in the vessels is modified. With a view also to lower the arterial pressure, certain agents have been recommended of late, of which the nitrites are the most efficient. Of these, nitroglycerine (or glonoin) has been the most extensively known and employed. Professor Rossbach (*Berlin. klin. Wochenschrift*, No. 3, 1885) has employed this remedy in a number of cases of renal cirrhosis with marked benefit. He says, "nitroglycerine is an excellent remedy in contracted kidney, well capable of removing threatening symptoms and of prolonging life." I have had some experience in the use of this remedy in these cases, and I am satisfied that it is one of great value. I have obtained not only marked reduction of arterial tension through its use, but also improvement in the character of the urine. Nitroglycerine is best administered in pill form, the usual strength of which is gt. j of the one per cent. solution to each pill. It is best to begin with one pill, and this may be repeated every two to five hours. It will usually be found that the dose can be increased considerably without disagreeable effects, after its use has been continued for a time. The nitrites of sodium and potassium in their action are very similar to nitroglycerine, lowering arterial tension through their vaso-dilator power over the arterioles. They are less likely to produce headache than nitroglycerine; and their action continues much longer. They have recently been employed for the purpose of lowering arterial tension in cases of renal cirrhosis with favorable results. Mr. R. M. Simon reports

(*Birmingham Medical Review*, Feb. 1885) the use of nitrite of sodium in five cases of granular kidney associated with increased arterial tension, in all of which improvement is claimed to have followed. The nitrite of sodium or potassium may be administered in aqueous solution in doses of three to five grains, and in order to secure continuous action the dose should be repeated every four or five hours. According to the observations of Dr. D. J. Leech (*British Medical Journal*, Nov. 28, 1885), cobalt yellow (potassio-cobaltic nitrite) reduces arterial tension for a much longer time than any of the other nitrites. He writes: "The alkaline nitrites produce but little influence on pressure for from six to ten minutes, but then depress it strongly for two and a half hours, and act altogether for between four and five hours. Cobalt yellow begins to act markedly in from twenty to thirty minutes; its tension-depressing effects are often very pronounced for more than three hours and sometimes visible for six hours." The doses of cobalt yellow employed by Dr. Leech were four to seven grains. Of its class, the nitrite of amyl acts as the most rapid and powerful vaso-dilator of the arterioles; but unfortunately its effects pass off so quickly that we are unable to secure any continued effect from its use. It has an immediate effect and continues to act for about two minutes, but its influence on the pulse subsides completely in from twenty minutes to half an hour. It should be borne in mind that the nitrites are equally applicable in the early stages of cirrhosis if there be marked increase of arterial tension; indeed, I believe their use under such circumstances tends to postpone cardiac hypertrophy.

The nitrites should be employed with caution if acute uræmia be threatening. I deem it the more important to urge this caution since certain of these agents are recommended for the relief of acute uræmia.

It is true that the nitrites diminish, to a marked extent, the sensibility of the reflex centres, hence their power sometimes to relieve convulsions; but other effects of these agents unfit them for use in cases in which active uræmia is threatening, as follows: (*a*) they usually diminish the secretion of urine; (*b*) they reduce to an extreme degree the oxidizing powers of the blood; (*c*) by their powerful vaso-dilator influence over the arterioles they permit the reflex centres to become unduly engorged with blood, thus greatly exposing them to the irritating influence of the waste products with which the blood is always overloaded in these cases.

I have seen acute uræmia precipitated in two cases by the administration of nitroglycerine.

Digitalis is contraindicated in the stage of cardiac hypertrophy prior to the development of dilatation, more especially if arterial tension be greatly exalted; in such cases its use may be attended by serious consequences. Iron, also, is unsuitable, and should not be employed.

The urine should receive the same attentive observation in this as in the early stage of the disease. It is less likely to become diminished in quantity, as the great pressure in the arterial system favors conditions of filtration. Accessions to the disease in the kidneys may, however, arise, which may cause oliguria, and this may be followed by uræmia. Diminution in the quantity of urine, increase of albumin, and the appearance of granular and epithelial casts, point to acute augmentation of the disease. The remedies to be employed are purgatives, the citrate and acetate of potassium, low diet, rest in bed, etc., until the exacerbation is relieved. The hot air bath, so efficient in these conditions in the early stage, must now be used with caution, or better altogether withheld, owing to its action in increasing arterial pressure.

With regard to the complications arising in this stage of the disease, hemorrhages are perhaps the most common. Should these occur within the cranium, we should not hesitate to bleed, apply ice to the head, and use brisk purgatives. Hemorrhages into the retina are to be met with less active measures, purgatives and ergot, to begin with, followed by small doses of mercury, or iodide of potassium. The remaining complications common to this stage of the disease are mostly of uræmic origin, and should be managed as advised in the chapter on uræmia. They are best avoided by maintaining the urine above normal in quantity, though in accomplishing this we must carefully avoid the use of those diuretics which increase the cardiac power, or act as vaso-constrictors on the arterioles, unless arterial tension be low.

ADVANCED STAGE.—Sooner or later the heart becomes unequal to the increased demands upon its power, and its muscular structure undergoes fatty changes and begins to weaken: the tension in the arterial system gradually becomes lowered; the urine diminishes in quantity; and the dangers become increased from nearly every source, save that of hemorrhages.

As soon as evidences of commencing failure in the circulatory forces become manifest, we should begin to support the cardiac muscle. Digitalis now becomes an agent for good in this disease. It should be given in small rather than large doses (five to seven drops of the tincture). To this drug should be added strychnia in small and repeated doses (one-fiftieth to one-thirtieth of a grain), continued over long periods of time. With a view, also, to nourish the cardiac muscle, a more liberal diet should be indulged in; at the same time the precaution should be observed of keeping the urine up to the normal in quantity. If the latter be attended to, we need not hesitate to allow the patient a moderate amount of meat with the principal meal of the day. It should be remembered that these patients are usually kept upon a rigid diet for months, perhaps for years, preceding the stage of cardiac dilatation, to the almost total exclusion of meats, and quite properly so. Now, however, the conditions are vastly changed, the dilating heart must be supported by every possible means ; otherwise, the cardiac failure will greatly complicate the case. The weakening of cardiac contractions permits the blood pressure in the tufts to become lowered, and filtration of urine is thereby diminished, entailing uræmic symptoms, and all those secondary complications which cut short the lives of so many of these patients. The necessity, therefore, of supporting the heart by every means available, should be constantly and prominently kept in view in the treatment of this stage of the disease.

Of scarcely less value than the agents just named, for the purpose of improving the cardiac tone, is the use of iron. This metal is indicated throughout the advanced stage of the disease. It may be administered in conjunction with the strychnia and digitalis.

The urine requires more careful watching in this than in any other stage of the disease. It is extremely liable to diminish in quantity, falling greatly below normal. It should, therefore, be measured daily, and the average noted from time to time—say, once a week. The solids of the urine, especially urea, also become diminished in the late stages of the disease. A sudden reduction in the gross quantity of the urine, more especially if accompanied by a marked diminution in the quantity of its solids, is often the precursor of acute uræmia. The potassium and sodium salts, combined with digitalis, convallaria majallis, and adonis vernalis, are among the most effective agents at our command for combating the failure of the renal function under such conditions.

I have obtained excellent results in a number of these cases from the administration of fluid extract of apocynum cannabinum in doses of five to twenty minims. It is an efficient diuretic and cardiac stimulant, but in large doses it is likely to induce purging.

Hot air baths are excellent adjuvants to the action of diuretics now, not only relieving congestion of the kidneys, but also acting as a cardiac stimulant.

If the quantity of urine falls below normal, and there remain any length of time, dropsy will become developed; and this may even become so extensive that measures must be employed for its relief. It owes its origin to the same influences which cause diminution of the excretion of urine, viz., failure of the cardiac power. It will now be necessary to call forth the cardiac forces by the use of the most effective agents at our command. Digitalis should be given in full doses—fifteen drops of the tincture, to which may be added five to ten drops of the fluid extract of adonis vernalis. The action of digitalis on the heart is sometimes remarkably increased, if given in conjunction with adonis vernalis. Convallaria majallis may be added to these, forming a compound which I have found exceedingly efficient in some cases. When the pulse slows down to a marked degree, as is likely to happen under its extended use, the dose should be diminished, or the mixture discontinued for a few days. It is well to remember that the cumulative effects of digitalis are relieved at once by the use of the nitrites. Since I have learned from experience that inhalations of nitrite of amyl afford instant relief from the over-action of digitalis and agents of its class, and that the relief may be continued under the use of nitro-glycerine, I now rarely fear the over-action of agents of the digitalis group upon the heart, if my patient be under daily observation. Strophanthus is said to act more rapidly and more powerfully upon the heart than digitalis, and it is, therefore, likely to prove a valuable remedy in the condition under consideration. Its dose is from two to six drops of the tincture.

If it be desirable to use more active eliminants in order to reduce the dropsical accumulation, nothing proves so effective as the administration of concentrated salines as previously described. These may often be given even when diarrhœa is present, without apparently aggravating the intestinal irritation; indeed, I have seen the diarrhœa improved after their administration.

The secondary inflammations arising in the last stage of the disease demand active measures for their relief; even under the most skilful management they often prove quickly fatal. As they are brought about by defective action of the kidneys permitting the blood to become loaded with waste products, which act as irritants to the inflamed tissues, the urine should be increased in quantity, and maintained above normal by the employment of appropriate measures already considered. Brisk purgatives are indicated also, in order to compensate for the defect of the renal function; in short, the general measures to be employed are the same as those given for the relief of uræmia, and need not be repeated here.

When inflammation of the lungs or pulmonary œdema arises —and these are among the most common secondary affections—I have obtained the best results from the administration of ergot combined with digitalis, but to be effective they must be employed early. The ergot may be given in doses of twenty to thirty minims of the fluid extract (preferably Squibb's), repeated every four hours if the stomach will tolerate it. Jaborandi must be avoided in pulmonary complications.

The retinal lesions must be treated according to their nature. If they are hemorrhagic—and such are less common in this stage—we must rely on leeching the temples at first; followed by the use of ammonium hydrochlorate, which is often very effective in removing the effused blood. It may be given in aqueous solution in doses of three to ten grains, and its taste is best disguised by the addition of extract or syrup of liquorice root. If the visual defect arise from pure retinitis, leeches and counter-irritants to the temples will give the best results. If the retinal lesion be degenerative in character—and this is the most common form now, with more or less retinitis in addition—we shall get the best results from the use of iron, quinine, strychnia, etc.; after first reducing any inflammatory action present by the appropriate measures just considered.

Acute uræmia is very common in this stage of the disease, and it is more often of the comatose than the convulsive form. It is probably the most frequent cause of death in the advanced stage of this disease. The treatment of acute anæmia will be found in detail in the chapter devoted to that special subject.

Diarrhœa occurring in the late stage of renal cirrhosis will often be found difficult to control. Opium, so effective in similar disorders from other causes, must be used with caution here; as these patients are peculiarly susceptible to the effects of opiates. It has long been known that opium is liable to produce serious, and even fatal consequences when administered to people whose kidneys are granular.

Very recently I had occasion to submit this question to a systematic test in a case of renal cirrhosis which was accompanied by amblyopia, cardiac changes, and the usual phenomena which characterize that disease. The patient had been under my care for over a year, and was in the late stage of the disease. Diarrhœa had proven very obstinate, although astringents had been employed with a view to check it. I resolved, at length, to resort to the cautious use of opiates for the purpose of modifying the diarrhœa. To prevent the decrease of the urinary secretion which I apprehended under the use of opiates, I first ordered a mixture of equal parts of tincture of digitalis, fluid extract of convallaria majallis, and adonis, vernalis, to be given in thirty drop doses, and repeated every six hours. I also ordered some pills, each to contain a half grain of powdered opium, and one-eighth of a grain of sulphate of copper.

The first day one pill was given after breakfast, and the urine for 24 hours measured 41 ounces, although for weeks previously it had measured from 45 to 60 ounces daily.

The medicines were repeated three days in succession, precisely as on the first day, with substantially the same result as to the quantity of urine passed. The fifth day one pill was given after breakfast, and one at bedtime— the digitalis mixture being continued as before—and the urine for 24 hours measured but 30 ounces.

The medicines were repeated on the sixth and seventh days in the same doses and frequency as on the fifth day, with substantially the same result as to the quantity of urine voided—which did not vary more than two or three ounces from thirty daily.

The opium was next discontinued for six days, the digitalis mixture being given as before ; and the quantity of urine gradually increased, until it reached 50 ounces in 24 hours.

The pills were then begun again, one being taken morning and evening, when the urine promptly decreased to 25 ounces in 24 hours. Next one pill was given daily, when the quantity of urine rose to 35 ounces in 24 hours. Efforts were thus made to reconcile the system to the presence of opium for over five weeks, during all of which time the digitalis mixture was continued regularly, while the opium was given interruptedly, and never exceeded one grain daily in half-grain doses eight hours apart. The result was practically the same as has been described, except that the system became more intolerant of the opium, until at length the quantity of urine fell to 20 ounces in 24 hours, and I was obliged to abandon the use of the opium on account of the appearance of uræmic symptoms. Nor was the use of the opium abandoned any too soon, for *after its discontinuance* the urine continued to decrease in quantity for four days, until it only measured 11 ounces in 24 hours. Marked uræmic symptoms followed—including uræmic intoxication —and it was only through active measures that I was enabled to reëstablish the urinary secretion, and save my patient's life.

It may also be stated that the number of stools passed by the patient daily when a grain of opium was given, averaged two or three; and that previous to the use of opium they had averaged from eight to twelve daily.

If the opium be administered to persons whose kidneys are cirrhotic, it not only diminishes the secretion of urine, but its narcotic effects become greatly prolonged, owing to the slow elimination of the drug by the kidneys; the pupils remaining contracted after a single dose, for from 24 to 28 hours or even longer. While, therefore, ordinary *single doses* may sometimes be given under such circumstances, without serious consequences, *repeated doses—*

especially at short intervals—are likely to lead to a cumulative action of the drug which may result in fatal narcotism.

It will be safer, therefore, to resort to other measures for the purpose of checking diarrhœa in the late stage of this disease. The vegetable astringents sometimes answer the purpose very well. Tannic or gallic acid combined with sulphate of copper forms a good combination for administration in such cases.

Finally, the importance of hopefully and persistently pushing onward with every judicious measure in the treatment of this disease, even in the face of the most discouraging and gloomy prospects, cannot be too urgently insisted upon. We possess no accurate means of determining during life the extent to which the kidneys may be damaged. If we conclude that because the urine continues for a time to be excreted in small quantity, that the kidneys are extremely atrophied, we may often be agreeably surprised to find the organs afterward prove themselves adequate to perform their function for months or even years.

Again, some of the complications of the disease are but efforts of the system to rid itself of the consequences of the disease, and, in this, nature may be at least temporarily successful. For instance, convulsive seizures may liberate in the circulation a large amount of urea ; and the well-known diuretic properties of the latter may stimulate the kidneys to renewed functional activity and thus the agencies that have been threatening the extinction of life may become modified, and the patient may survive without further accident for years.

CHAPTER VI.

SCARLATINAL NEPHRITIS.

ACUTE nephritis, well marked, and possessing special characteristics, accompanies or follows a large number of cases of scarlatina. The proportion varies greatly in different epidemics; in some it ensues only exceptionally, while in others it occurs in a very fatal form in more than half of the cases. It is, therefore, difficult to arrive at precise conclusions as to the average proportion of cases of scarlatina that are followed by nephritis, but it has been roughly estimated at about one in six. Thus, it will be seen that the number of cases of nephritis arising from scarlatina is enormous, probably greater than from any other cause, and amply entitles it to separate and careful consideration.

The above estimate by no means includes all cases of albuminuria that arise during the course of scarlatina. Both Dr. Hillier and Dr. Dickinson found the urine to contain albumin in half the cases of scarlatina treated in the Children's Hospital in London.

While I do not claim that every case of albuminuria arising in the course of scarlatina is due to actual nephritis, yet I shall endeavor to show, from clinical and pathological facts, that a very large number of the cases of pyrexial albuminuria, so-called, occurring in scarlatina, are distinctly nephritic, and accompanied by the usual lesions of nephritis. Dr. Henoch (*Lectures on Diseases of Children*) correctly writes, "a further development of the disease from such a beginning is by no means rare; and, on the other hand, nephritis may be found at the autopsy, although no albumin was present in the urine during life."

Unfortunately, as yet, we possess no reliable data by means of which we are able to predict, with any degree of certainty, the particular cases of scarlatina most likely to be accompanied or followed by nephritis. It has been claimed by Wagner and Leichtenstein, that if the cervical glands become secondarily enlarged after once subsiding, nephritis is especially likely to accompany or to follow scarlatina. Others have claimed, that in cases in which nephritis becomes engrafted upon scarlatina, the fever does not subside at the usual period of apyrexia (about the sixth day), but that the temperature remains exalted through the second, and perhaps the third week. But it will be remembered that the cervical glands are often intensely inflamed in cases in which no nephritis accompanies or follows scarlatina; and, moreover, the continuation of pyrexia in scarlatina is more often due to other sequelæ of the disease than to nephritis, and, therefore, cervical adenitis and continued pyrexia cannot be considered reliable indications of the presence of nephritis in these cases.

Neither does the violence of the scarlatina itself necessarily imply more danger of nephritis in these cases, for it is well known that some of the very mildest cases of scarlatina—so mild that they may be overlooked—are followed by most fatal types of nephritis; while, on the other hand, in some of the most severe cases of scarlet fever the kidneys escape damage.

The late Dr. Mahomed insisted that scarlatinal nephritis is preceded by increased arterial tension, constipation, and the presence of the blood crystalloids in the urine, the latter being determined by means of the guaiacum test. Should these statements be confirmed by future observations, they will afford valuable evidence of the approach of nephritis in scarlatina.

ETIOLOGY.

PREDISPOSING CAUSES.—The most prominent predisposing cause of scarlatinal nephritis appears to be the age of the patient. Scarlatina itself is less frequent during the first than subsequent years up to the fifth. Murchison's statistics of 148,829 deaths from scarlatina, in England and Wales, give a total of 9999 cases as occurring in the first year of life, or about 1 in 15 of the whole. *Scarlatinal nephritis* is rare during the first year—in fact, very few cases have been recorded. After this age, however, it becomes more and more frequent up to the latter part of the third year. About the beginning of the fifth year of life its frequency commences to decline, and it becomes less and less common till the twentieth year of age, after which, like scarlatina, it is rare.

Scarlatinal nephritis occurs more frequently in the male than in the female sex, in the proportion of about three to two.

The stage of scarlet fever has much influence in the production of nephritis. By far the larger number of cases of nephritis arise, or, to speak more accurately, become manifest for the first time during the latter part of the third week, or from the eighteenth to the twenty-fourth day. Less often, nephritis occurs during the second week, and less frequently still in the fourth, fifth, or sixth week. If nephritis be first discovered later than the sixth week, it has probably existed from an earlier date and escaped detection.

Family predisposition probably exercises some influence in the causation of this, as it does in that of most other forms of nephritis. Thus several observers have noted instances where nephritis complicated nearly every case of scarlatina occurring in certain families. Lastly, some peculiarity in the character of certain scarlatinal epidemics—the nature of which is at present unknown—very markedly favors the development of the nephritic complication.

EXCITING CAUSES.—It was long held that exposure to cold during the desquamative period of scarlatina was the exciting cause of this disease. From the intimate relationship known to exist between the skin and kidneys, and the fact that acute nephritis is a common result of exposure to cold, it was natural to suppose that the desquamation occurring in scarlatina rendered the skin peculiarly susceptible to the influence of cold, and thus the latter more readily excited the nephritis.

Circumstances, as well as facts, however, strongly controvert the above supposition. In the first place, as was shown in the third chapter of this work, exposure to cold rarely gives rise to acute nephritis in children under any circumstances. In the second place, as we shall presently see, the anatomical changes in the kidneys resulting from scarlatinal nephritis are altogether different from those of nephritis resulting from exposure to cold. Lastly, if nephritis of scarlatina is excited through exposure to cold, then it should be more frequent among the poorer classes where exposure is vastly more common than among the rich; but statistics show that such is not the case.

It is at present generally held that the exciting cause of this disease is the specific poison of scarlatina itself, and that it is eliminated by the kidneys, exciting inflammation in those organs in its passage through them.

As to the nature of the scarlatinal poison, nothing positive is yet known. Numerous observers have asserted that the blood of scarlatinous patients contains micrococci in abundance, and there are at least two microbes described as giving rise to scarlatina—the *monas scarlatinosum* of Klebs, and the *plox scindeus* of Eklund. The supposition, however, that nephritis of scarlatina is caused by microorganisms rests at present more largely on analogy than upon facts; and as we are still ignorant of the nature of the scarlatinous poison, its further consideration would be purely speculative.

MORBID ANATOMY.

The naked-eye appearances of the kidneys in scarlatinal nephritis vary according to the length of time the disease has been in progress when death occurred. Within the first ten days or two weeks these will be found to vary but little from the normal. The kidneys usually contain more blood than normal, and therefore look congested, but the size of the organs remains unchanged.

Later in the disease, and continuing up to the third month, or longer, the organs will be found enlarged and usually firmer to the touch than normal. Their surfaces present to the eye a mottled appearance, caused by admixture of small yellowish spots with the natural color of the organs. The organs may appear paler than normal.

In very late stages of the disease the swelling of the organs will be found to have mostly subsided, and the kidneys, in some cases, may be found smaller in size than normal. The kidneys are much harder to the touch than in health, and they present increased resistance to the knife when cut.

The microscopic characters of the kidneys will be found to vary quite as much as their external appearances, according to the length of time the disease has existed. For the purpose of rendering these more clear, I shall describe the usual changes found in each stage separately.

MORBID ANATOMY. 197

First.—When the disease proves fatal within the first week, as it sometimes does from convulsions, suppression of the urine, etc., the following changes are observable under the microscope:

Scattered through the cortex in various locations are seen aggregations, or circumscribed masses of corpuscles, and within, or very near these, the vessels are enlarged

FIG. 16.

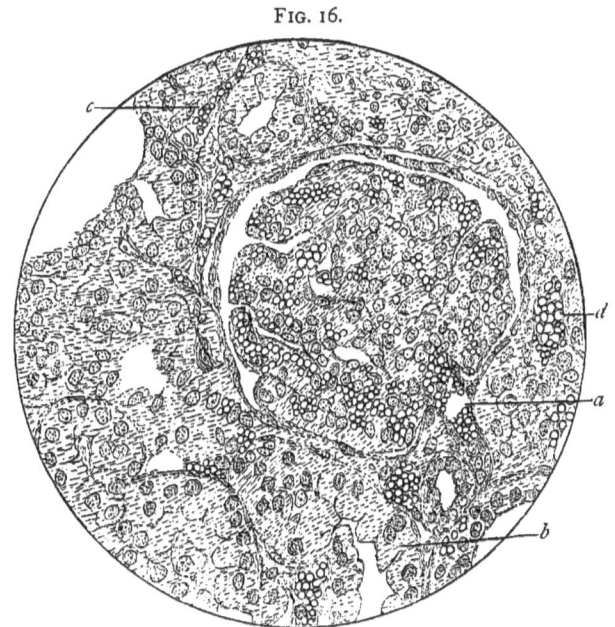

Scarlatinal nephritis, fourth day., Cortical section. *a.* Intracapsular hemorrhage passing into tubule. *b.* Cloudy swelling of tubular epithelium. *c.* Intertubular hemorrhage. *d.* Intratubular hemorrhage.

and crowded with corpuscles. The situations of these masses of corpuscles are both inter- and intratubular, as well as glomerular, but they are always found most typical within and without the capsule of Bowman. Between the vessels of the tuft these corpuscles are especially numerous, while external to the tuft and within the capsule of Bowman collections of considerable extent may often be seen, in some cases passing into the tubes when the latter are clear.

In some cases the vessels of the tuft are seen to be ruptured, giving rise to free hemorrhage, which may fill the capsule of Bowman. The epithelium lining the convoluted tubes is in the condition known as cloudy swelling. The cells are much enlarged and project into the tubules, in places almost closing them. This is best seen in cross-sections of the tubes. The outlines of the epithelial cells are more angular than normal, and the nuclei are obscured. The contents of the cells are granular in appearance. The epithelium lining the capsule of Bowman partakes of the same changes as that of the convoluted tubes. The diameters of the convoluted tubes are distinctly increased, as are those of the glomeruli. The interlobular arteries are prominent and thickened, as are also the arterioles. This thickening, according to Klein, is due to hyaline swelling of their intima, and infiltration of their muscular coat with nuclei.

Second.—When the disease proves fatal between the second and sixth weeks the following changes are observed in addition to those just described. The glomeruli are much swollen and surrounded by considerable granular-looking matter. The capsule of Bowman is thickened, and hyaline in appearance, and contains numerous small pink nuclei. The tufts may be either swollen and turgid, filling the capsule of Bowman, or they may be pushed aside from extravasations of blood, or desquamated epithelium, or both. The connective tissue corpuscles between the vessels of the tuft are increased in number. The epithelium lining the capsule of Bowman is seen to be rapidly undergoing desquamation, large collections of exfoliated epithelium being detached and lying in the space between the tuft and the capsule of Bowman. See Fig. 17.

The epithelium of the convoluted tubes is granular and fatty, and, like that lining Bowman's capsule, it is desqua-

FIG. 17.

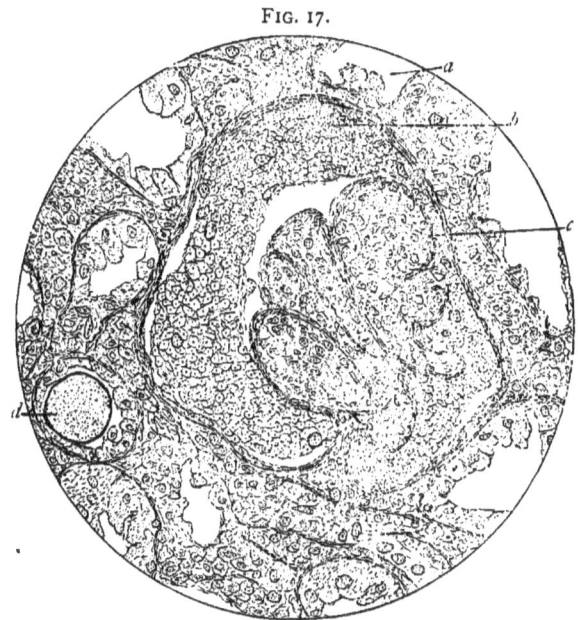

Scarlatinal nephritis, third week. *a*. Tubule from which epithelium has been exfoliated. *b*. Desquamated epithelium of Bowman's capsule.. *c*. Compressed tuft. *d*. Cast in tube seen in cross-section.

mating. In some places the tubes are seen to be choked with granular and fatty débris, or with colloid or mixed casts; while, in others, they may be found comparatively clear.

The thickening of the interlobular arteries and arterioles is more pronounced than in the condition last described.

Third.—When the disease proves fatal after the second month the following changes will be found:

The capsule of Bowman is greatly thickened, and its outer margin is ill defined. The sharp outlines of the capillary tuft are lost in the increased nucleated cell proliferation between its vessels; in fact, the whole Malpighian body is greatly swollen and presents the appearance of a dense, homogeneous, granular mass. Great increase of connective tissue corpuscles is observed surrounding the capsule of Bowman, more especially near the exit of the efferent vessel. Between the convoluted tubes, particularly about the efferent arterioles, may be seen patches of dense tissue, containing new connective tissue corpuscles. The convoluted tubes are variously affected. Those nearest the glomeruli are often seen to be compressed by surrounding inflammatory growth; while those midway between the glomeruli retain their normal outlines; but the epithelium lining the tubes is flattened, and the lumens of the tubes, which may be larger than normal, usually contain more or less granular débris. Colloid casts, both in the convoluted and straight tubes, may usually be seen in this stage.

The interlobular arteries and arterioles are thickened, in some cases greatly so, and along the terminal branches of the former may often be seen wedge-shaped patches of dense-looking tissue, containing new connective tissue corpuscles, just as is observed in the early stages of cirrhosis of the kidney.

It will be seen that the anatomical changes in the disease under consideration are most marked in and immediately surrounding the glomeruli in all stages, and hence they have been described by Klebs under the name of "glomerulo-nephritis." It should be borne in mind, however, that cases are met with in which the anatomical changes are more diffuse from the beginning, the whole tubular tract of the cortex sharing in the changes progressing in and about the glomeruli.

It may be pertinently asked here, Do the lesions in the kidneys consequent on scarlatina arise in the beginning of that disease, and pursue a latent course for a longer or shorter time before they are disclosed? Recent investigations seem to give an affirmative answer to this question. Steiner found that in scarlatinal patients who died early the kidneys were *always* diseased. In children who died on the second and third days from this disease, though he found the kidneys hyperæmic and enlarged, with cloudy swelling of the renal epithelium, the urine presented no evidence of disease.

Thomas says (Ziemssen's *Cyclopædia*, vol. ii.) that at the beginning of the disease the urine frequently shows a mucous cloud, and cylindroidal casts; albuminuria occurs only exceptionally at this period.

Klein has found, during the first week of scarlatina, a hyaline change in the Malpighian tufts, evidences of disturbed function in the capsular epithelium, and nuclear infiltration of the muscular coat of the small arteries.

Dr. R. Stevenson Thomson has recently communicated (November 10, 1885) to the Royal Medical and Chirurgical Society the results of his examination of the urine of 180 cases of scarlet fever. Special attention was directed to the early days of scarlatina, and 35,000 specimens of urine were examined from the 180 cases. The frequency of mild and evanescent attacks of nephritis in scarlatina was brought out very distinctly. The evidence seemed to favor the view "that nephritis was a feature of scarlet fever almost as essential as the rash or sore throat."

I have in my own collection a section of the kidney

taken from a child who died from scarlatina at the thirty-sixth hour of that disease in which all the changes represented in Fig. 16 are clearly presented.

SYMPTOMS.

The disease usually begins with pyrexia. In mild cases this may be so slight that it is scarcely observable; while in the severe type of nephritis the fever is considerable. This pyrexia is secondary to that arising at the onset of scarlatina, and during the second stage usually subsides.

For some days preceding the appearance of fever, the child usually exhibits evidences of uneasiness or actual discomfort: the appetite is lost, headache is complained of, and constipation may be present.

With the pyrexia the pulse becomes accelerated, reaching 100 to 120; the tongue becomes slightly coated, and nausea or vomiting may occur. The urine now becomes diminished in quantity, its specific gravity is reduced, and it usually contains albumin, casts, and blood. Œdema soon makes its appearance in the feet or face, and extends with a rapidity proportionate to the severity and extent of the nephritis. A noticeable pallor creeps over the countenance of the child, which becomes more marked with the progress of the disease.

It should be borne in mind that cases occur in which all the symptoms just spoken of are so greatly modified, that they escape observation until attention is directed to them by the appearance of dropsy.

The Urine.—This varies greatly with the stage and activity of the disease. In the beginning the deviations from normal are usually slight, and consist chiefly in an increase of mucin, the presence of hyaline casts, and albumin, the latter often transient in appearance. Sooner or later, however, the urine becomes reduced, sometimes to 200 or 100 centimetres, or it may even be suppressed; more often it is passed in larger quantities, though it is always more scanty than normal. It is usually acid in reaction, and its specific gravity is reduced, averaging 1.012 or thereabout. The color varies with the stage of the disease: it may be light yellow, reddish-brown, or black in color; the two latter being due to the presence of blood, and usually present during the more active course of the disease; while the lighter colors are more often present during the latter stages. The urine is usually "smoky" in appearance, owing to the presence of undissolved urates and organic débris; and this character is more marked the more scanty the urine becomes, and hence it is most frequently observed during the acute stage of the disease. In the late stages of the disease, the urine usually clears, often becoming transparent. The urine contains casts. The appearance of these in the urine is among the earliest evidences of the disease; and they usually continue some weeks after albuminuria has subsided. At first the casts are mostly hyaline; later epithelial and bloody casts are present, characteristic of the acute stage; still later, the granular and fatty casts predominate, though the fatty variety is less numerous than in other forms of nephritis, because there is less fatty degeneration of the renal epithelium in scarlatinal than in other forms of nephritis, save perhaps the puerperal. The urine usually contains a number of red blood disks, especially in the more acute period of the disease. These may be so numerous as to give the urine a reddish-brown color. If the urine is very acid in reaction, and the corpuscles are very numerous, the color will be dark (blackish). The urine contains white blood corpuscles in varying numbers, broken-down epithelium, and granular débris, consisting of urates, oxalates, and disorganized cells. Free crystals of uric acid are often present, as are, also, epithelial cells which have become detached entire, or in part, from the lining of the convoluted tubes and capsules of Bowman.

Albumin is usually present in the urine, but varies in quantity very much, probably more than in any other form of nephritis. In the earliest period of the disease it may be present during only part of the day. During the height of the disease the quantity of albumin in the urine is usually considerable, and it is often excessive. The quantity may be large one day, and much diminished the next. The quantity may become gradually increased in some cases, while in others it is subject to sudden and great increase in the course of a few hours.

Lastly, it should be borne in mind that exceptional cases occur in which the urine is free from albumin throughout the disease. Some authors question the occurrence of the above condition, but so many observers have noted it, especially Simon, Becquerel, Phillips, Rilliet, Barthez, Rayer, Frerichs, Steiner, and more recently Thomas, Gee, Henoch, Bartels, Dickinson, Smith, and Goodhart, that no further doubts remain of its occasional occurrence.

Goodhart has recently described minutely several cases of the kind which occurred in his own practice (*Diseases of Children*, p. 83, 1885), and devotes several pages to the consideration of " dropsy without albuminuria," resulting from scarlatina.

Dropsy.—This is a scarcely less fitful symptom of scarlatinal nephritis than is albuminuria. It may develop slowly, and never become pronounced; or it may occur suddenly and extend rapidly, reaching an extreme degree in the course of a few days. After the dropsy has attained considerable headway, it may remain with little apparent fluctuations for weeks or months, or it may subside in a few weeks or days. The anasarca usually corresponds inversely with the quantity of urine excreted; and the higher grades of dropsy indicate extensive damage going on in the kidneys. In the milder grades of dropsy the abdominal cavity only is invaded; while in the more pronounced form the pleural or pericardial cavity may become the seat of serous effusions. Hydrothorax is especially to be dreaded in this disease, owing to the marked tendency of the effused fluid to assume a puriform character, which adds greatly to the gravity of the case. Finally, dropsy may be absent during the whole course of scarlatinal nephritis, but in such cases the disease is always mild.

The Digestive System.—As already stated, the appetite is usually lost early in the disease, and it generally remains more or less impaired throughout. Cases occur, however, in which the appetite remains good, even during the acute period of the disease; and this may mislead the physician as to the gravity of the case, unless he be on his guard.

Vomiting is one of the most frequent symptoms of scarlatinal nephritis. This may be purely reflex, or it may be of uræmic origin. In the former case Henoch points out (op. cit.) that other symptoms, viz., headache and somnolence, which give a serious significance to "uræmic" vomiting, are absent. It may also be stated, that purely reflex vomiting is most common early in the disease, before the latter has become fully developed; while the vomiting of uræmic origin is mostly present after the disease has become well established.

The bowels are usually constipated in scarlatinal nephritis, more especially during the pyrexial stage. Though this is undoubtedly the rule, diarrhœa is not an infrequent exception after the disease has continued one or two weeks. Sometimes the stools are profuse and watery, and accompanied by severe pain or cramps. This diarrhœa has been regarded as an "eliminative effort of nature" to rid the system of the specific virus of the disease; but it is more likely due to the excretion by the intestinal mucous membrane of urea, which, undergoing decomposition in the intestines, liberates carbonate of ammonium, which is very irritable to mucous surfaces.

The Circulatory System.—A remarkable and striking series of cardiac disturbances occur in a large number of cases of scarlatinal nephritis. The proportion of these remains as yet to be accurately determined; but from the statistics available I feel justified in placing the proportion of cases complicated by cardiac derangements, in one form or another, as at least one-third of the whole. The most frequent form of cardiac change resulting from scarlatinal nephritis is undoubtedly that of acute hypertrophy. If the chest be examined physically in every case of scarlatinal nephritis, it is remarkable in how large a number of the cases the apex beat of the heart will be found displaced downward and to the left, accompanied by accentuation of the second sound. This hypertrophy of the left ventricle of the heart sometimes becomes developed in an almost incredibly short period of time—frequently a few days are sufficient to bring about marked changes. As evidence of the rapidity of these changes it may be noted that the cardiac impulse is often found to be altered in its location in the course of two or three consecutive days. If the nephritis of scarlatina terminates favorably within six or eight weeks, the cardiac hypertrophy usually passes away, in some cases almost as quickly as it appeared.

In a more limited number of cases of scarlatinal nephritis the heart undergoes acute dilatation. This is an extremely grave complication, as it is so liable to cause œdema of the lungs. Cardiac dilatation in these cases is usually announced by a frequent, thready, fluttering pulse, cold extremities, and frequent respirations. Dilatation of the heart due to scarlatinal nephritis is different in its nature from that which so frequently occurs in the late stage of renal cirrhosis. In the latter cases it arises in consequence of degeneration of the cardiac muscle, and is, therefore, permanent. In the case of scarlatinal nephritis no fatty changes in the heart are discoverable at the autopsy, and if the nephritis terminate favorably the cardiac dilatation may be completely recovered from.

It is altogether likely that both hypertrophy and dilatation of the heart in scarlatinal nephritis are due to some influence which calls greatly on the cardiac power. If the heart be able to respond to the increased demands made upon it, hypertrophy of the left ventricle is the result only. If, on the other hand, the cardiac muscle be unequal to the extra strain, dilatation of the ventricle is the result. Still another form of cardiac derangement is quite frequent in scarlatinal nephritis. This is manifested by symptoms of collapse ; the pulse slows down to sixty-five, perhaps even to fifty beats in the minute, and often becomes alarmingly irregular. The respirations become quickened, the extremities cool, and the same danger of pulmonary œdema is incurred that arises from dilatation of the heart. As no constant cardiac lesions are present in this form of derangement, it is possible that some influence arises which paralyzes the cardiac muscle. Henoch suggests (op. cit.) that the symptoms are due to disturbances in the innervation of the pneumogastric nerve from uræmic influences. He has seen a number of cases in which these "collapse-like" disturbances of the heart's action were associated with distinct signs of uræmia.

Pericarditis and endocarditis occasionally arise in the course of scarlatinal nephritis. They pursue a rapid course and almost invariably prove fatal, as they do when they complicate other forms of nephritis.

From what has been said in reference to cardiac hypertrophy in this disease, it will be readily understood how exalted vascular pressure is often present in scarlatinal nephritis, more especially during the early stage. Indeed, the late Dr. Mahomed insisted that high arterial tension preceded nephritis in all cases of scarlatina, and that it marked what he termed the "prealbuminuric stage" of the disease. Although many have disagreed with that lamented observer as to the causes of high arterial pressure in these cases, yet most admit that it is a valuable evidence of approaching danger. We may have increased tension of the pulse before nephritis is discoverable by other symptoms. It may accompany the early course of nephritis, or it may continue to the end if hypertrophy of the heart be constant. If, however, cardiac dilatation takes place at any time, the pulse at once loses its tension and becomes dicrotic.

Lastly, anæmia quickly becomes a marked feature of scarlatinal, as of other forms of acute nephritis.

The Respiratory System.—The consequences of scarlatinal nephritis here are very similar to those resulting from acute nephritis from other causes. A greater liability to pulmonary œdema is encountered in the scarlatinal form, owing to the occasional occurrence of cardiac dilatation in the latter. In such cases, also, the œdema is usually more rapid in development than in nephritis from other causes. The respirations quickly increase from forty to sixty in the minute, and the lungs fill up so rapidly with serous infiltration that death may result in a few hours. Pneumonia and pleurisy are frequent causes of death in this disease. Bronchial catarrh is not infrequent; and as long as it remains uncomplicated is not of serious consequence; but it is readily kindled into the acute form, which may extend to the pulmonary tissue, and thus prove fatal.

The Nervous System.—Acute uræmia is the most prominent consequence of the disease referable to the nervous system. This occurs with about the same frequency in scarlatinal nephritis that it does in acute nephritis from other causes, pregnancy excepted. It may be stated, however, that uræmia less frequently proves fatal in scarlatinal than in any other form of nephritis. Henoch has observed (op. cit.) that the nephritis of scarlatina is recovered from even more rapidly than usual after uræmia has occurred. I believe this to be equally true of other forms of nephritis that are curable, provided the convulsive seizures are survived.

Visual disturbances of all kinds are probably less frequent in scarlatinal nephritis than in any other variety of acute renal disease. Amblyopia of the pure uræmic type, without obvious retinal lesions, occasionally occurs. The impairment or loss of vision is sudden in its development in these cases, and usually vision is restored in the course of a few days. Rarely hemorrhagic retinitis occurs, the vision remaining impaired for weeks.

Œdema of the brain, serous effusion beneath the cranium (hydrocephalus externus), or, more rarely, into the ventricles (hydrops ventriculorum), are occasional results of this disease. These may be a part of the general anasarcous condition of the patient, or some of them may arise in consequence of violent convulsive seizures.

Various other disturbances of the nervous system are occasionally observed, of which headache is the most common. Occasionally, acute delirium occurs in these cases, immediately preceding or following the convulsions. Tinnitus aurium, deafness, insomnia, restlessness, and mental confusion, are sometimes present. In themselves they are not of serious consequence, and are usually transient; but their presence should be regarded as ominous of a coming uræmic storm.

Partial or local paralyses, and even hemiplegia, have been recorded as following convulsions in scarlatinal nephritis. These are extremely rare occurrences, however, and are probably always recovered from if the patient survive the other consequences of the disease.

COURSE AND DURATION.—In the majority of cases this disease is first recognized during the third week, less often during the second or fourth week. The first symptoms usually are moderate but increasing albuminuria, mingled with casts and blood. The urine is diminished considerably in quantity, and is passed more frequently than normal. Dropsy soon follows, or it may even be the first symptom noticed. It first appears in the face (in the eyelids or cheeks), less often in the feet, unless the child be about. The dropsy extends till a general anasarcous condition, more or less marked, is the result. If the disease pursue a favorable course, and no complications arise, improvement will begin after a week or ten days. The quantity of urine is increased to normal or even considerably beyond, the quantity of albumin is decreased, the blood disappears from the urine, or decreases so greatly in quantity that it is no longer apparent to the naked eye. The dropsy now begins to decline, and if the symptoms continue to improve, albumin usually disappears in the course of three or four weeks and recovery results.

While the above is the course of the disease in most cases, yet it does not always pursue such an uncomplicated career. The disease exceptionally sets in with suppression of the urine, which is quickly followed by convulsions or other acute uræmic manifestations. In other cases such extreme symptoms are reached more gradually—after several days or weeks. Relapses may occur at almost any period of the disease: the case may be proceeding favorably up to the third or fourth week, or even later, when suddenly the urine may decrease in quantity, becoming bloody and highly albuminous; dropsy may extend quickly, involving the chest, and the life of the patient may be placed in extreme danger from pulmonary œdema or uræmia.

In other instances the case may apparently be progressing favorably, when suddenly the pulse may become feeble, irregular, and sometimes slow. The respirations may become rapid and the extremities cool, and death may result suddenly from collapse.

In other cases, the disease, though perhaps uncomplicated, pursues an obstinate course. The dropsy does not yield, and albumin continues to be excreted in large quantity. The disease may continue thus for weeks or months and pass into the chronic form.

In a considerable number of cases the physician is first called to treat the child for dropsy; no definite history of scarlatina is ascertainable. Perhaps a vague account of transient indisposition may be elicited as having occurred many weeks previously. In such cases the disease has passed into the chronic form and may continue indefinitely.

Finally, in a limited number of cases—and I desire to

emphasize this fact—the patient is brought to the physician, suffering from cardiac hypertrophy with slight albuminuria, or it may be that his vision has been failing, and this is found on examination to be due to "albuminuric retinitis." A history of scarlatina, with dropsy six, eight, or even ten years previous, is elicited. Such is an example of the cases which have passed from under treatment before nephritis has been cured, the renal disease continuing in a latent form for years, till at length, when discovered, the case has entered the list of hopeless ones, and life has become irretrievably blighted in its hours of promise. It is to be feared that this sad picture—which is no rare one—is too often the result of failure to continue the treatment until all evidences of the disease have disappeared.

From what has been said, it will be readily understood, that both the course and duration of scarlatinal nephritis are subject to great variation. The disease may pursue an uninterruptedly mild and favorable course from the beginning; its onset may be so violent that death may result in a few hours; or it may assume a grave form at any time during its course, after having previously progressed auspiciously.

If the disease prove fatal in its acute stages, death usually occurs within the first eight or ten weeks. On the other hand, if the disease pursues a very mild course, or if it passes into the chronic form, it may continue for several years—two to fifteen.

DIAGNOSIS.

CONSTRUCTIVE.—In typical cases of scarlatinal nephritis the renal inflammation is readily recognized. The same characteristics which reveal the presence of acute nephritis from other causes may be relied upon to disclose the presence of nephritis of scarlatinous origin. These have been fully dwelt upon in the chapter on acute nephritis.

It is only in those rare cases of scarlatinal nephritis, in which the disease pursues a latent or irregular course, that special difficulties in recognizing the disease arise, and these demand attentive consideration here.

In every case of scarlatina, no matter how mild, the urine of the patient should be examined daily, and this rule should be followed especially during the second, third, and fourth weeks from the onset of the disease. The examination should comprise not only a search for albumin, but the microscope should be repeatedly appealed to for evidences of the presence of renal disease, for it is now known that scarlatinal nephritis may be present, while albumin is absent from the urine, at least temporarily so.

If nephritis be present, even in the mildest possible form, the urine will contain renal epithelium, casts, and usually some blood corpuscles. The casts may be cylindroidal, hyaline, or epithelial. The epithelium will be cloudy, granular, and more or less degenerated or partly broken down, showing its disturbed nutrition previous to desquamation. It is true that if only a small portion of the urine be examined, in some cases none of these products may be found; but this does not prove their absence from the urine. It is only after having collected the whole urinary sediment of twenty-four hours, and submitting a part of it to the microscope that we are able to say positively that these products are absent from the urine; and it is only under similar circumstances that we are safe in assuming that nephritis is, or is not present, in doubtful cases.

DIFFERENTIAL.—In well-marked cases of the disease occurring at the usual periods of scarlatina, if the latter be recognized it will be almost impossible to mistake the form of renal disease present. In those cases, however, in which scarlatina has been overlooked in consequence of its mild character, doubts may arise as to the form of the renal lesion. Dropsy and other evidences of nephritis may be present, but no very definite history of recent illness is obtainable. In such cases careful inquiry as to the recent appearance of scarlatina in the family or neighborhood, and a critical examination of the patient for evidences of recent desquamation, especially on the fingers or toes, or a history of recent sore throat, often throw sufficient light upon the question to lead to a correct diagnosis of the form of renal lesion present.

The characteristics which distinguish acute nephritis in general from other forms of renal disease have been considered in the third chapter of this work, and they are practically identical with those of scarlatinal nephritis.

PROGNOSIS.

The prognosis in general in scarlatinal nephritis is extremely uncertain. While probably two-thirds of those attacked by the disease recover, yet its course is so capricious that a guarded prognosis should always be given. More cases of scarlatinal nephritis terminate in recovery than those of acute nephritis from most other causes; probably, in part, because the renal structure undergoes less extensive fatty changes in scarlatinal nephritis; and in part because the reparative powers of the system are greater in the young subjects of scarlatinal nephritis than in older subjects of acute nephritis from other causes. But the disease is so likely to be attended by accidents, and relapses are so frequent that we are never justified in rendering an unqualifiedly favorable prognosis in any case.

The favorable, as well as the unfavorable indications, are practically the same as those in acute nephritis from other causes. The only important exception to this is the occasional development of cardiac complications, especially cardiac failure in scarlatinal nephritis, which must always be regarded as extremely grave.

TREATMENT.

PROPHYLACTIC. —If precautionary measures were always adopted in the treatment of scarlatina with the view to prevent nephritis, it is probable that the renal complication would occur less frequently, and that many of the more violent cases of scarlatinal nephritis would become considerably modified. In all cases of scarlet fever it is, therefore, highly essential that appropriate measures should be employed from the beginning, for the purpose of preventing or modifying the renal inflammation.

Perhaps the most important of these preventive measures consist in a careful regulation of the diet. While the same general principles should be observed, which have been advised, in the first chapter, as appropriate in the management of albuminuria, yet certain facts and features connected with scarlatinal nephritis render a special consideration of this subject appropriate in this place.

The most striking circumstance in this connection is the fact already mentioned, that scarlatinal nephritis is of rare occurrence during the first year of life. The only circumstance that seems adequate to account for this rarity of scarlatinal nephritis during infancy, is the fact that these patients subsist exclusively upon milk. We learn from this the important lesson that an exclusive milk diet is the most efficient prophylactic against scarlatinal nephritis at our command. The child should, therefore, be subjected to an exclusive milk diet, from the commencement of scarlet fever, and this should be continued to the end of the fourth week.

While it is not desirable to purge the patient actively, yet the bowels should not be permitted to become constipated, as the latter condition is exceedingly liable to raise the blood pressure in the vessels and precipitate the renal inflammation. The late Dr. Mahomed held that the failure of the bowels to move but for a single day, was often sufficient to bring about the above results. To children under four years of age a grain of hydrarg. cum. cret. for each year of the age, may be given at bedtime, if the bowels do not move during the day; and should this not produce a movement of the bowels by the following morning, its action should be aided by means of an enema. The hygienic measures laid down in the first chapter, for albuminuria in general, should be carefully carried out. All violent exertion should be guarded against—indeed, not only should the little patient be confined to bed, but his whims, as well as his wants, should receive such attention as is calculated to prevent violent or long-continued fits of crying. Finally, scarlet fever should never be treated by the internal administration of certain drugs, more especially chlorate of potassium and the salicylates. This caution would scarcely be requisite, were it not for the fact that some authorities still advise the use of these drugs in scarlatina. It may, therefore, be stated that both chlorate of potassium and the salicyl compounds are renal irritants, capable of giving rise to albuminuria and nephritis when administered internally in doses generally considered harmless.

CURATIVE.—It may be unhesitatingly asserted that digitalis is the most powerful weapon we possess against the nephritis consequent upon scarlatina.

If, at the autopsy, however, we look at the engorged renal glomeruli, infiltrated and surrounded by exuded blood corpuscles, in those cases in which death has occurred during the acute stages of scarlatinal nephritis, we might perhaps naturally be led to hesitate in administering such agents as increase the power of the heart, and consequently the force of the blood current. But digitalis does something more than this. It acts as a powerful vaso-constrictor to the arterioles, and, therefore, to the vessels of the tufts. Fothergill, Gourvat, and Ackermann long since demonstrated this by the action of the drug on the web of a frog and mesentery of the rabbit under the microscope. While, therefore, the blood pressure is increased under the influence of digitalis, and consequently filtration of the urine is hastened; in addition to this the calibre of the vessels of the tufts is diminished, the permeability of their walls is decreased, and corpuscles cease to exude. The desquamated epithelium within the capsule of Bowman, clogging the entrance to the convoluted tubes, is forced out by the increased filtration from the tufts; the flow of urine is reëstablished; the hemorrhage is checked; the inflammatory products are washed away, and the inflammation itself is controlled. Clinical experience has amply demonstrated the superior efficiency of this drug in the treatment of these cases; indeed, before its true *modus operandi* was clearly understood it had won its way into popular favor as a remedy for scarlatinal dropsy; and more recent and thorough knowledge of its pharmacological action has shown that its beneficial results in such cases are due to its power of controlling the renal inflammation upon which the dropsy depends.

Digitalis should, therefore, be given as soon as the renal inflammation is discovered, and its administration continued as long as acute symptoms persist. It may be given in the form of tincture in doses of one to five drops, according to the age of the child, and it should be repeated every sixth hour.

If the nephritis be mild in character, as evidenced by slight albuminuria, absence of blood from the urine, and the quantity of the urine remaining little reduced, digitalis should still be given in moderate doses. In such cases it acts as a preventive against the onset of more acute grades of nephritis, by narrowing the pathways of the blood through the terminal branches of the interlobular arteries and vessels of the tufts. In addition to this, it sustains the heart, retards cardiac hypertrophy, and often prevents cardiac dilatation and collapse.

Should the symptoms assume a more acute form, as shown by the quantity of urine becoming markedly decreased, and especially by hæmaturesis becoming prominent, digitalis should be given in larger doses. A very advantageous method of its administration, under such circumstances, is to give three or four full *medicinal* doses at six hours intervals, and then to fall back to medium doses again. By this means we test the susceptibility of the patient to the action of the drug without incurring the risks of its cumulative effects. Should the urine become increased in quantity, and the blood diminished by this means, these favorable changes may often be caused to continue under the smaller doses of the drug.

Ergot, also, has been used in the acute grades of this

disease, principally with a view to check the discharge of blood from the kidneys. It sometimes effects this purpose, but with much less certainty and efficiency than digitalis, and in several other respects it is greatly inferior to the latter in this disease.

The anuria of scarlatinal nephritis differs in its causes to some extent from that which occurs in other forms of acute nephritis. When acute nephritis from cold, etc., was under consideration, it was stated that anuria is chiefly due to blocking up of the convoluted tubes with colloid matter. While this may be the case to some extent in scarlatinal nephritis, also, yet it is altogether probable that anuria in the disease under consideration is more largely due to changes within the glomeruli, notably the profuse desquamation of the intracapsular epithelium. It is well known that this desquamation of epithelium from the lining of Bowman's capsules in most cases of acute scarlatinal nephritis is enormous, in many cases blocking up the intracapsular spaces and entrances to the convoluted tubes. It has already been shown in Chapter III. that if the urine be rendered alkaline, the free colloid matter occupying the convoluted tubes becomes dissolved and washed out with the urine. Now *free* colloid matter wherever found obeys the same chemical laws, and is sooner or later dissolved when brought into contact with alkaline solutions. If, therefore, the urine be rendered alkaline in these cases, it attacks the protoplasm of the necrosed and partly broken down epithelial cells within the capsules of Bowman rendering their complete disorganization and solution more rapid, and greatly facilitating the clearing away of the obstructions on which anuria largely depends.

For the purpose of rendering the urine alkaline, the salts of potassium formed with vegetable acids are the most desirable preparations. They should not be given on an empty stomach, but during digestion, as they then effect the purpose in view more rapidly and effectually.

The citrate of potassium is, on the whole, perhaps the most desirable preparation for administration to children in these cases, owing to its less disagreeable taste. It may be given in doses ranging from two to ten grains, according to the age of the child. It should be moderately diluted with water, and it may be repeated every two to six hours, according to the effect produced on the reaction of the urine. I have not observed the deleterious effects of these preparations of potassium on the system which are said to arise sometimes from their internal administration, notably cardiac weakness and diminution of arterial blood pressure, although I have given the citrate and acetate of potassium in considerable doses,—just short of the purging point—uninterruptedly for months.

I am inclined, therefore, to believe that it is chiefly those salts of potassium which do not part with their oxygen in the system, and are excreted undecomposed, that produce the injurious effects referred to. Bartholow has pointed out that this is especially the case with the nitrate and chlorate of potassium. This much is certain, that the vegetable salts of potassium are destroyed in the system, and are eliminated as alkaline carbonates (possibly excepting the bitartrate, Wood), and in all probability in the latter form they are comparatively harmless, at least in medicinal doses, even to young children.

The salts of sodium formed with vegetable acids render the urine alkaline, in the same manner as do those of the potassium base. They do not produce the deleterious effects on the organism attributed to the potassium salts just considered, but they are less efficient than the latter.

As it is sometimes necessary to maintain alkalinity of the urine uninterruptedly for considerable time in these cases, the salts of sodium may be occasionally substituted for those of potassium. For this purpose the acetate of sodium may be given in solution in doses of two to ten grains, according to the age of the child (adult dose, fifteen to thirty grains). A more pleasant form for administration is the effervescent citro-tartrate of sodium (B. P.), which may be given in water, in doses ranging from five to thirty grains (adult dose, ʒj to ʒij). But the granular effervescent citrate of potassium and citro-tartrate of sodium, as prepared by Bishop, of London, are the most palatable forms in which to administer these salts; which is an important consideration in the case of young children.

Diarrhœa, which is not uncommon, should be modified, but rarely checked completely, unless great exhaustion results therefrom, for the reasons already explained. Threatening constipation, on the other hand, should be averted by the use of mild aperients, aided, if necessary, by enemata.

General bloodletting has been found very efficient in many cases in relieving the more acute symptoms. If the child be of robust constitution, and the symptoms do not yield to other measures, from two to six ounces of blood may be taken from the arm, often with the happiest results. The application of leeches to the lumbar region will greatly assist in overcoming the more active symptoms of this disease in many cases. Cupping the loins, so often recommended, is inapplicable in the case of young children. Hot fomentations to the loins are excellent measures, as they are in other forms of acute nephritis, and their use should not be neglected. If the more acute symptoms become somewhat modified, while the renal inflammation evinces a lingering tendency, counter-irritation to the lumbar region is often followed by prompt improvement in all the symptoms of the disease. For the purpose of effecting this, croton oil is the safest and most efficient agent to employ. It may be mixed with olive oil in the proportion of one part of the former to two, three, or four parts of the latter, according to the age of the child. I prefer to make these applications to the sides, in the concavities immediately above the crests of the iliac bones, where they are equally efficacious as when applied direct to the lumbar region. An advantage of the former locations is, that the child is not caused unnecessary suffering when lying on his back; moreover, the irritated cutaneous surfaces are out of the line of pressure, even when the child lies on his side.

Quinine has been highly extolled in the treatment of scarlatinal nephritis. In cases in which serous effusion has commenced in the pleural cavity quinine is undoubtedly serviceable, as it tends to prevent the effused fluid assuming a purulent character. Aside from the above complications, quinine—especially in large doses, as is recommended—is of doubtful utility in this disease, since large doses of this drug enfeeble the action of the heart, and diminish the excretion of urea and uric acid by the kidneys.

Attention should be given to cardiac complications when they arise in the course of scarlatinal nephritis. If the area of cardiac dulness is found to be extending, the impulse of the apex beat to be changing its location, and the second sound of the heart to be more sharply accentuated than normal, we should resort without delay to cardiac supportives, as these symptoms indicate that hypertrophic changes are in progress.

Digitalis, if not already being given, should now be administered in moderately full doses. If, notwithstanding this, the symptoms of hypertrophy of the heart become more marked, the action of digitalis may be aided by the administration of convallaria majallis. The fluid extract of convallaria majallis may be given in conjunction with digitalis in doses of three to ten drops. The extract of the leaf only should be used, as but little of the properties which act upon the heart reside in the root of the plant, and, moreover, the latter is emeto-cathartic to an active extent (Bartholow). The convallaria may be given in full doses and continued indefinitely, as it is not cumulative in its action. In many cases the progressing cardiac hypertrophy will be stayed by resort to the above measures.

Less frequently, hypertrophic changes in the heart continue unchecked and pass into dilatation of the ventricle, or the latter may occur suddenly ; or the pulse may become irregular or slow with symptoms of collapse, as previously remarked. We have then to deal with an exceedingly grave complication.

The extended observations of Dr. Thomas R. Fraser on the action of strophanthus (see *British Medical Journal*, Nov. 14, 1885), lead us to hope that in this drug we possess a more efficient remedy for these special cases than in any of the digitalis group hitherto employed. According to Dr. Fraser's investigations, strophanthus is a more powerful and rapid stimulant to the cardiac muscle than is digitalis, and it is, therefore, peculiarly suited to the cases under consideration. The adult dose of the tincture of strophanthus is from two to six drops, and Dr. Fraser has found no evidences of cumulative action of the drug from its long-continued administration.

Although strophanthus is invaluable in cases of pure cardiac failure, it should not be substituted for digitalis in the *general treatment* of scarlatinal nephritis. Unlike digitalis, strophanthus exerts little or no influence over the arterioles, its whole power over the circulatory apparatus, according to Dr. Fraser, being expended upon the heart. It will, therefore, be understood that in those cases in which the nephritis is very acute, and the urine very bloody, the continued use of strophanthus would be likely to aggravate the *renal* mischief. Under these circumstances digitalis outranks all other drugs in point of efficiency in subduing renal inflammation.

The range of utility of strophanthus in scarlatinal nephritis is limited to cases characterized by great weakness of the heart's action, and in such cases it promises to prove the most efficient cardiac stimulant at our command.

The remaining symptoms, as well as complications, of scarlatinal nephritis, are for the most part identical with those occurring in the course of acute nephritis from other causes, and the measures designed for their relief demand no special consideration here, as they have been detailed in Chapter III.

Before leaving this subject, it only remains to insist upon the great importance of treating all cases of scarlatinal nephritis, however mild, until all evidences of the disease shall have passed away. The absence of albumin from the urine does not constitute sufficient evidence of restoration to dismiss the case permanently from observation. Casts are usually to be found in the urine of these patients, two, three, or four weeks after albuminuria has subsided, and in some cases considerably longer. These teach us that nephritis still lingers in some portion of the renal tract in a latent form, or, at least, that the structure of the gland has not completely regained its normal condition.

So long as casts are present in the urine the patient

should be kept under at least occasional observation. It is, indeed, advisable that the urine should be carefully examined, both chemically and microscopically, several weeks after all traces of nephritis have disappeared from the urine.

When the medicinal treatment of scarlatinal nephritis is discontinued, special instructions should be given to observe a careful regulation of the hygiene of albuminuria (Chapter I.) for several months.

CHAPTER VII.

PUERPERAL NEPHRITIS.

THE occurrence of albuminuria during the period of gestation attracted no special attention from early writers. It was not until Rayer published his *Traité des Maladies des Reins*, in 1839, that this subject began to attract notice. Rayer wrote that he frequently observed that his "*nephrite albumineuse*" occurred in women who were pregnant. His pupil, Cohn, asserted later that albuminuria and convulsions of pregnancy were due to nephritis. We now know that both albuminuria and nephritis occur in a large number of cases of pregnancy, and that convulsions of a very fatal type are liable to accompany them.

The existence of albuminuria during gestation does not always imply that nephritis is present; for slight albuminuria is not uncommon during the gravid state when the kidneys are not inflamed.

The reports of lying-in hospitals give the proportion of cases of nephritis occurring during gestation as 1 in 130 to 1 in 150. It was formerly held that puerperal nephritis is confined mostly to the latter months of gestation, but more recent observations have shown that the disease arises usually during the middle months.

Statistics show that the majority of cases of puerperal nephritis occur for the first time among primiparæ. It has, therefore, been assumed that primiparæ are more liable to the disease than multiparæ, but this does not necessarily follow. It would be safer to state, as Dr. Ralfe

ETIOLOGY. 227

writes (op. cit.), "that women who have not been the subject of nephritis in their first pregnancy are less liable to the disease in subsequent pregnancies."

ETIOLOGY.

Lever was probably the first to claim that puerperal nephritis is due to the pressure exerted by the gravid uterus upon the renal veins. Rosenstein, Simpson, and others advocated the same view; and, for some time, this explanation was accepted without dissent. Recently, however, the mechanical theory of the production of puerperal nephritis has met with opposition, especially from Bartels, on anatomical grounds.

The following facts seem to invalidate the theory of the mechanical causes of puerperal nephritis:

(*a*) As the gravid uterus rises from the pelvis, it does so in the line of the axis of the pelvic inlet, which is directed away from the vertebral column, and impinges against the anterior abdominal wall. Now, as the lower end of the uterus is within the pelvic cavity, it follows, as Bartels has remarked, that "the uterus would have to be bent backward on itself at a considerable angle just above the inlet to the true pelvis in order to be able to touch the anterior surface of the second lumbar vertebra" (the location of the renal veins).

(*b*) If we even admit that the gravid uterus is capable of so changing its normal position as to impinge upon the second lumbar vertebra, it could only compress the left renal vein as it crosses in front of the abdominal aorta; the right renal vein passes out from the kidney to the vena cava across a space which is protected from direct pressure of the uterus in front by the vertebral column. If, therefore, *direct* pressure of the gravid uterus were the cause of the renal disturbance, the left kidney only should become affected; while, as a matter of fact, in puerperal nephritis both kidneys are simultaneously affected.

(*c*) That puerperal nephritis does not arise from *general* abdominal pressure is shown by the fact that intra-abdominal pressure due to other causes than pregnancy, as from large growths of the ovary, or even of the uterus itself, does not cause nephritis. The ascites sometimes associated with abdominal growths is not, as has been claimed, due to nephritis, or even to pressure. M. Ferrier has shown before the Société de Chirurgie that these growths are accompanied by ascites only when vegetations are developed on their exterior. " The secretions discharged into the peritoneum by these pseudo-glands constitute a colloid material which sets up in the peritoneum osmosis, whence arises the constant increase of peritoneal fluid—that is, ascites." Moreover, Méhee has pointed out that this fluid contains more solids than the ascitic fluid of cardiac or renal disease—66 to 71 grammes, instead of 59 per kilogramme of fluid.

(*d*) Neither the anatomical changes in the kidneys, nor the symptoms of puerperal nephritis correspond to those produced by venous stasis from other causes.

(*e*) Puerperal nephritis frequently arises during the early months of gestation, before mechanical pressure can be taken into consideration as an exciting cause.

These facts seem to render the mechanical theory of the production of puerperal nephritis untenable.

To what, then, are we to ascribe the nephritis which so frequently arises during the course of gestation? On the whole, it seems most reasonable to suppose that puerperal nephritis is brought about by some change in the blood. Frerichs pointed out that the pregnant woman is unusually hydræmic; that the blood contains less albumin and red corpuscles than normal; and that the white corpuscles are increased in number. While this may have a predisposing influence, yet it seems likely that some disturbing element in addition is present in the blood, which acts upon the renal structure as an irritant; for a hydræmic condition is often present to an extreme degree in many states of the system besides that of pregnancy without our being able to recognize it alone as a common cause of nephritis.

It has been alleged that the direct cause of puerperal nephritis is a toxic one, consisting of the waste products of tissue metamorphosis going on in the fœtal organism, and that these 'passing through the maternal blood, are eliminated by the kidneys, causing irritation and sometimes inflammation in those organs during their excretion. At least one circumstance seems to support this view, namely, that in twin, or multiple pregnancy, the tendency to nephritis is much greater than when gestation is single. This was so apparent to Litzmann that in doubtful cases he decided against twin gestation when the urine contained no albumin (Bartels).

Acute atrophy of the liver owes its origin with great frequency to the puerperal state, and Braun and Heller have observed that the blood drawn during pregnancy frequently contains the bile elements in considerable quantities. Dr. Fehling, in a memoir recently read before a meeting of German scientists at Strasburg, called attention to the frequency of habitual death of the embryo in puerperal nephritis, in cases in which there is no evidence of syphilis.

It is not improbable, therefore, that during the puerperal state some disturbing element is present in the blood which acts as an irritant to the kidneys, and which is sometimes capable of exciting inflammation in those organs, as well as acting injuriously upon the organism in other ways. We must, however, admit that we are ignorant of the exact nature of the cause of puerperal nephritis, as also of the cause of scarlatinal nephritis; and for the present these questions must be left to be determined by future investigation.

MORBID ANATOMY.

Perhaps in no other form of nephritis do the changes in the kidneys vary so much, both under the microscope and to the naked eye, as in puerperal nephritis. This fact probably accounts for the widely different opinions held by various authors as to the pathology of the disease under consideration. If death results from eclampsia during or near labor at term, the kidneys are usually found enlarged, and their consistence altered, the organs feeling less firm and resistant than normal. The surfaces of the organs are smooth, and their capsules non-adherent. On section, it is noted that the cortex is deeply injected with blood, and the cut surfaces present a brownish-red appearance The pyramids are hyperæmic and striped.

On microscopic examination it is observed that the cortex is the part most affected. The epithelial cells lining the convoluted tubes are seen to be swollen, and in some places they have become fatty. Hemorrhagic extravasations are less frequent and pronounced than in other forms of acute nephritis, especially the scarlatinal form; yet patches of migrated blood corpuscles may sometimes be seen between the convoluted tubes, and about the glomeruli. Colloid matter is observed, sometimes in abundance, blocking up the convoluted tubes; with this, fatty and epithelial products are sometimes mixed. The interlobular vessels are distended, as are also the afferent arterioles. The glomeruli are enlarged in consequence of the distention of the vessels of the tufts. In some cases the cortex is less vascular than that described.

If, on the other hand, the patient should survive the period of labor, and nephritis pass on into the chronic form, or should succeeding pregnancies be accompanied by repeated attacks of nephritis, we may find after death altogether different anatomical changes in the kidneys. In such cases the appearances of the organs often approach those of chronic nephritis, even showing slight atrophy, with granular surfaces and slightly adherent capsules; in short, the changes are those of chronic nephritis in its late stages, described in Chapter IV.

In the early course of puerperal nephritis the anatomical changes differ from those in other forms of nephritis, chiefly in their less pronounced degree; the more intense grades usually occur only after considerable duration of the disease. The anatomical changes found in the late stages of puerperal nephritis differ from those of chronic nephritis from other causes, chiefly in the more pronounced connective tissue changes in the kidneys found in the former.

SYMPTOMS.

Puerperal nephritis usually begins without very active symptoms, and it may so continue for weeks or months, before dropsy, eclampsia, or other marked symptom directs attention to the kidneys. The general condition of the patient often remains unimpaired, and there is so little to attract attention that the disease may remain undiscovered until the later months of gestation are reached; indeed, there is reason to believe that, in some cases, the disease continues through the period of labor, and afterward subsides without having been discovered. More often, however, some symptom, most frequently dropsy, directs attention to the disease at an earlier period.

Dropsy.—While this is usually first observed in the feet and lower extremities, it is not limited to these locations in puerperal nephritis. It also appears in the face, and, sooner or later, in other parts of the body, and this fact serves to distinguish the dropsy of puerperal nephritis from the œdema of the limbs which is not uncommon in pregnant women whose kidneys are free from disease. Dropsy of puerperal nephritis is usually more tardy in its development than that associated with acute nephritis from other causes. Exceptionally, cases are met with in which dropsy reaches a comparatively high grade in the course of a few days.

The extent of the dropsy depending upon puerperal nephritis is subject to considerable variation. Sometimes it merely amounts to moderate œdema of the subcutaneous areolar tissues, and even that may be confined to certain locations. It may continue throughout in this mild form, with but little fluctuation. In other cases anasarca becomes so extreme that it is difficult for the patient to walk about. If the nephritis remains untreated, dropsy usually increases up to the termination of labor, after which it begins to decline, and in favorable cases quickly passes away. Although ascites is quite common in puerperal nephritis, the thoracic cavity is less frequently the seat of serous effusion than in acute nephritis due to other causes.

Lastly, in a considerable number of cases of puerperal nephritis no dropsy whatever occurs, and in such cases the nephritis must be considered as very mild in character.

The Urine.—As a rule, the urine becomes diminished in quantity in puerperal nephritis. The decrease may be so slight at first that it is not noticed, but actual measurement will, in nearly every case, demonstrate that the quantity is below the normal. For weeks or months this falling off may be but moderate or slight. At any time, however, the urine is liable to sudden decrease, though it seldom is completely suppressed. The degree of diminution may be taken as a fairly reliable index of the extent and gravity of the nephritis in individual cases.

The specific gravity of the urine is usually inversely as the quantity of urine excreted, it being higher when the quantity passed is reduced, and *vice versa*. It is seldom, however, that the urine in puerperal nephritis attains the high specific gravity seen in acute nephritis from other causes. It may range slightly above normal (1.022 to 1.028), or it may descend below normal (1.015). After the termination of labor, if other symptoms improve, it is not uncommon for the specific gravity to fall to 1.012.

The color of the urine is subject to considerable variations. During the more active grades of the disease it is usually dark in color, yet it seldom contains many red blood-corpuscles, as in other forms of acute nephritis; neither is it as turbid or smoky as in the latter, although it is not clear, especially when the specific gravity is high. The urine contains hyaline, granular, and fatty casts; and renal epithelium is usually present. It may be stated, however, that the urinary sediment is usually much less in quantity, as a whole, in this than in any other form of acute nephritis. The percentage of urea, as a rule, is reduced, though subject to great variation. Albumin is usually present, often in remarkably large quantities. The latter fact has been especially spoken of by numerous observers, and has been held to be due, in a large measure, to the altered condition of the blood common in pregnant women. However that may be, the fact remains that in these cases the quantity of albumin present is greater in proportion to the actual extent of the renal lesions than in any other form of nephritis. At first the quantity may not be excessive, especially if nephritis arises during the early months of gestation. In the later months, however, it is mostly always large in quantity, especially during or immediately preceding eclampsia.

The Digestive System.—In the early course of the disease the stomach often escapes disturbance, which seems remarkable when we reflect that the reflex irritability of the stomach is often greatly exalted during gestation. As the disease advances, however, nausea is very likely to be present, especially if the disease is of a pronounced type. The derangement of the stomach due to nephritis should not be confounded with that arising from pregnancy. In the former nausea and vomiting are likely to occur at any time during the day, and the appetite is usually lost; while in the latter the nausea and vomiting are mostly confined to the early hours of the day, and the appetite often remains unimpaired The bowels are less frequently disordered in puerperal nephritis than in other forms of acute nephritis.

The Circulatory System.—The pulse is usually slow, full, incompressible, and marked by high tension, as shown by the sphygmograph. Anæmia may be noticeable, but usually to a much less extent than in other forms of acute renal diseases.

I have been particularly impressed with the frequent occurrence of cardiac complications in chronic nephritis of puerperal origin. The majority of cases which have come to me in which nephritis either continued through or reappeared in succeeding gestations, were associated with cardiac complications. I do not believe this was due to preëxisting cirrhosis of the kidneys, as most of these cases occurred in patients under thirty years of age.

SYMPTOMS. 235

The usual form of cardiac lesion in such cases is hypertrophy of the left ventricle, which, I believe, runs a more rapid course than does cardiac hypertrophy associated with genuine cirrhosis of the kidneys. My reason for this belief is the fact that I have observed dilatation of the heart to follow upon cardiac hypertrophy in these cases at a comparatively early date. In one case in a woman aged twenty-seven years, nephritis being present in two consecutive gestations, in each of which abortion occurred about the fifth month, very marked dilatation of the heart developed and proved fatal within eighteen months from date of first gestation.

I am satisfied that cardiac complications are much more frequent in the late stages of puerperal nephritis, than in chronic renal inflammation from other causes, primary cirrhosis of the kidney excepted.

The Nervous System.—Of all the consequences of puerperal nephritis, none is so obvious as its effect upon the nervous system; indeed, in no other form of renal disease is the nervous system so frequently and so profoundly disturbed.

If we take into consideration the fact that the reflex susceptibilities of the nervous system are highly exalted in the pregnant state, we shall probably be led to a correct solution of the cause of this frequent disturbance of the nervous system in the disease under consideration. Certainly no other circumstance seems adequate to account for it. The renal lesions are rarely so profound in puerperal nephritis as in acute nephritis from other causes; in fact, they are often remarkably slight in the former. Neither is there anything special in connection with the renal function itself which should cause us to apprehend greater damage to the nervous system in these cases. The quantity of the urine and its percentage of urea do not usually suffer greater reduction in puerperal than in acute nephritis from most other causes. It therefore seems reasonable to conclude that the pregnant state strongly predisposes the nervous system to the disturbing influences of nephritis; and this conclusion is in harmony with our knowledge of other disturbances of the system consequent upon gestation.

The most prominent disturbance of the nervous system resulting from puerperal nephritis is uræmic convulsions, more commonly known in this connection under the name of eclampsia. In no other form of acute nephritis is this complication so frequent or so fatal; and for these reasons it demands careful consideration.

It is not claimed that all convulsions occurring during pregnancy and labor are uræmic. There are no reasons why epilepsy, hysteria, etc., should not arise in the puerperal as well as in the non-puerperal state; in fact, from what has been already said as to heightened reflex susceptibility of the nervous system during pregnancy, it will appear evident that both hysterical and epileptic convulsions are even more likely to occur in the pregnant than in the non-pregnant condition. While the above is quite true, yet both clinical observation and the anatomical changes present in the kidneys in the great majority of the more severe cases of puerperal convulsions, show that they are of renal origin, and uræmic in character.

Although puerperal convulsions usually arise during labor, they may occur as early as the third month of gestation, or be postponed until after delivery. The statistics of Braun and Weiger show that more than one-half of the cases of puerperal eclampsia occur during labor; of the remaining cases, about twenty-four per cent. occur before labor, and about the same percentage after delivery. If, however, we take into consideration the brief period of labor compared with the whole period of gestation, including the puerperal state, the preponderating influence of labor in the production of eclampsia will become more apparent. This fact should not be lost sight of when the question arises as to the propriety of inducing premature labor in any given case.

The most frequent form of visual disturbance in this disease is "uræmic amaurosis." This develops suddenly, causing great impairment, or complete loss of sight. In most cases it arises previous to labor, and is sometimes the first symptom that attracts attention and leads to the discovery of nephritis. In most cases it quickly subsides after the termination of labor. Amaurosis from degeneration of the optic nerve (albuminuric retinitis) occurs much less frequently. The cases in which I have observed this condition have been chiefly those in which nephritis has either continued through pregnancy and become chronic, or recurred in successive gestations.

Headache is said to be more severe in puerperal than in most other forms of nephritis. It is especially likely to precede for some time the outbreak of convulsions; hence, when present, it serves as a warning of the approach of eclampsia.

Lastly, Litzmann has observed the frequent appearance of neuralgic pains in the region of the kidneys in puerperal nephritis. As these attacks were invariably accompanied by a diminution of the urine, which latter became restored in quantity when the pain subsided, Bartels suggested that temporary compression of the ureter by the gravid uterus might account for these symptoms.

The remaining symptoms of puerperal nephritis are similar in most respects to those of acute nephritis from other causes. With the exception of convulsions, the secondary complications of puerperal nephritis are comparatively uncommon. When, however, they do arise, they are essentially the same in their character and manifestations as the complications of acute nephritis from other causes.

COURSE AND DURATION.—Puerperal nephritis usually pursues a mild course for several weeks or months; and in such cases it is in reality a subacute form of nephritis. There is usually a slight diminution in the quantity of urine, and the latter does not contain blood in noticeable quantity. The appetite may remain good, dropsy be absent, and the patient feel as well as usual. Nephritis may arise during the third, fourth, or fifth month of gestation, and pursue this mild course up to within a few weeks or days before the commencement of labor. In most cases, however, dropsy appears during the latter months of gestation. It is most marked in the lower extremities, but appears also in the face and upper extremities.

With the development of dropsy the urine diminishes more markedly in quantity and becomes highly albuminous. The appetite is impaired or lost; and headache is liable to be severe. At any time during the course of such symptoms the patient may be attacked by convulsions, which are *always* preceded by considerable decrease in the excretion of urine. Dr. W. L. Richardson, of Boston, has published (*Gynecological Transactions*, vol. iii., 1879) the results of his observations upon this point in twenty-six cases of puerperal nephritis. In six of these cases the quantity of urine became gradually diminished, until the amount secreted in twenty-four hours was very small, and then began attacks of eclampsia. These ceased, or grew less frequent, as the urinary secretion became reestablished. In five cases the daily quantity of urine, although diminished, did not fall nearly as low as in the first class of cases; and although the general symptoms were more marked than in any of the cases in which convulsions occurred, yet the labor was normal and there were no attacks of eclampsia. In a recent note to the author, Dr. Richardson writes that he has observed forty cases of puerperal nephritis up to date, all of which were confirmatory of the above course.

Most frequently, convulsions are postponed until some time during labor, though other symptoms may be threatening. Sometimes, eclamptic attacks do not arise till after labor is terminated, generally before the cessation of the after-pains; though they may occur several days after delivery.

The patient may succumb at any time during eclamptic seizures. Usually, however, death does not result until the convulsive seizures have been repeated several times; the intervals between the attacks having been occupied by increasing coma and insensibility. The patient may die either during an eclamptic seizure, or in the comatose state which follows. Puerperal nephritis sometimes pursues a more active course from the beginning. Although the urine rarely contains blood, yet dropsy, albuminuria, and other symptoms are marked. In such cases death of the fœtus followed by abortion is liable to occur, especially about the fifth month of gestation.

In other cases the nephritis pursues a mild course for several weeks or months, when suddenly the symptoms become greatly aggravated. In some cases the patient may be seized with attacks of eclampsia without active preceding symptoms, even dropsy being absent. In such cases it will usually be found that the quantity of urine has undergone marked decrease preceding the eclampsia.

Puerperal nephritis quickly subsides after the termination of labor, if the patient survive the complications of the disease. The exceptions to this rule consist mostly of those cases in which nephritis has preëxisted—as during a preceding pregnancy,

When the disease continues and becomes chronic, atrophic changes in the kidneys usually follow, with most of the symptoms usually present in renal cirrhosis.

The duration of puerperal nephritis, except in cases in which the disease becomes chronic, is more definite than in any other form of nephritis. In most cases it does not extend beyond six months from its beginning. When, however, the disease continues into the chronic form, its duration becomes uncertain. The patient may succumb within the first year or two from some of the complications of the disease. If these be survived, life may be prolonged for four or five years, or longer.

I am led to conclude from observation of a number of these cases, that chronic nephritis consequent to pregnancy pursues a more rapid course than does chronic nephritis arising from most other causes.

DIAGNOSIS.

CONSTRUCTIVE.— Puerperal nephritis would seldom escape recognition if the practice were established of frequently examining the urine of the pregnant woman after the third month of gestation. The urine usually contains albumin, even when the disease is very mild. It has been claimed that both nephritis and eclampsia have occurred when no albumin was present in the urine. It has never been conclusively proven, however, that eclampsia has occurred in a case where albumin has been absent from the urine continuously for a considerable length of time preceding such accident (Cazeaux).

On the other hand, albumin is sometimes present in the urine of pregnant women whose kidneys are not inflamed. In the latter case albuminuria is slight, while in puerperal nephritis it is usually considerable, and often excessive.

The urine in cases of puerperal nephritis probably always contains renal casts. These are chiefly hyaline, granular, and epithelial; fatty casts being rare except in chronic cases. The urine also contains more or less renal epithelium, which has become granular and partly disorganized.

When the symptoms of puerperal nephritis are well marked, it is almost impossible to overlook the existence of the disease. In such cases the occurrence of headache, nausea, and vomiting, and more especially diminished secretion of urine, anasarca, and a high degree of albuminuria, should at once lead to the discovery of nephritis.

DIFFERENTIAL.—In distinguishing puerperal from other forms of nephritis, the first essential point consists in diagnosticating the pregnancy. Careful inquiry should next be made as to evidences of the existence of nephritis previous to conception. If after having diagnosticated pregnancy we are able to determine that nephritis was absent before conception, we may consider it probable that the nephritis is of puerperal origin. If, in addition to this, the urine is highly albuminous, but free from blood, and the general condition of the patient remains good, we are justified in assuming that the case is one of puerperal nephritis.

PROGNOSIS.

Should the patient escape the more grave complications of the disease, such as eclampsia, and should no previous renal disease have been present, the prognosis of puerperal nephritis may be considered good.

Should nephritis have existed during a previous gestation the prognosis will be less favorable; for while many of these cases recover, yet a number pass into a chronic and incurable form. As a general rule, the earlier in the course of gestation nephritis arises, the more unfavorable the prognosis, unless abortion occurs.

The immediate prognosis in any given case will depend much upon the urgency of the symptoms. If the quantity of urine becomes greatly reduced, especially without corresponding increase of its specific gravity, the prognosis must always be considered grave so long as this continues. If, on the contrary, the quantity of urine is but slightly reduced, and its specific gravity remains normal or above, the case presents a favorable phase, for eclampsia seldom arises under such circumstances.

The greatest danger to be apprehended in puerperal nephritis is the occurrence of attacks of eclampsia. The period of gestation at which this complication occurs, to some extent determines the degree of danger incurred by its appearance. Thus, if the convulsions arise at any time anterior to the commencement of labor, they may be regarded as more unfavorable than if they do not arise until after delivery. The character and frequency of the eclamptic attacks, also, very largely determine the degree of danger consequent upon this complication. Thus, if the convulsions be violent, frequent, long continued, and associated with marked coma, they must be looked upon as extremely unfavorable; while, on the other hand, if the convulsions be mild in character, of short duration, with lengthening intervals between the attacks, and the patient retains her consciousness, they may be regarded more favorably.

Lastly, the prognosis will depend, to some extent, upon the length of time the disease may have existed before its discovery. Cases that are recognized early and subjected to appropriate treatment, usually pursue a mild course, escape serious complications, and end in recovery. But if, on the contrary, the disease remains long undiscovered, and be not recognized until the appearance of uræmic symptoms, as vomiting, severe headache, amblyopia, etc., the outlook will be more gloomy.

TREATMENT.

In the majority of cases, puerperal nephritis demands less active treatment than does acute nephritis due to most other causes; for, as already shown, the renal inflammation is usually comparatively mild.

In ordinary uncomplicated cases we may rely more upon hygienic measures, and less upon active medication, than in acute nephritis from most other causes.

Nephritis consequent to pregnancy presents a peculiarly promising field for the employment of a milk diet. In all cases, therefore, in which milk agrees with the patient, it should be made use of to the exclusion of other food. Tarnier's treatment, which is said to have proven very successful in these cases, consists chiefly of an absolute milk diet, in conjunction with baths at a temperature of from 98° to 100° F., repeated every third or fourth day. Drinking a tumbler of hot milk while in the bath or immediately after, greatly promotes its diaphoretic action. When milk disagrees with the patient, it is important to enforce a non-nitrogenous diet, owing to the great danger of uræmia.

Special attention should always be given to the daily quantity of urine secreted in cases of puerperal nephritis; and every effort should· be made to prevent its decrease to a marked degree below normal. The percentage of the urinary solids should frequently be carefully noted (see rule in Chapter II.), and compared with the quantity of urine secreted.

Should the quantity of urine become reduced, it will be necessary to resort to the use of diuretics. The most suitable agents of this class are the acetate and citrate of potassium, digitalis, convallaria majallis, adonis vernalis, strophanthus, and scoparius.

If the disease assume an unusually acute form, the best results will follow the use of the citrate or acetate of potassium (in sufficient doses to maintain the urine alkaline), conjoined with digitalis.

Should the urine remain free from blood, scoparius or apocynum may be administered in the form of infusion or fluid extract. They often cause prompt increase of the urinary secretion and improvement in the other symptoms of the disease. Arterial tension should be maintained above normal by the use of cardiac stimulants, to favor filtration of urine from the vascular coils within the glomeruli. In acute cases digitalis is the most useful agent of this class; but if the urine be free from blood, any agent of the digitalis group, as convallaria, adonis vernalis, or stophanthus, may be employed.

The importance of maintaining the functions of elimination in an active state, should be kept prominently in view throughout the treatment of this disease, for by such means the danger of eclamptic attacks is greatly diminished; more especially if the renal function become seriously impaired, should compensatory elimination be actively stimulated. Purgatives, hot air baths, etc., should therefore be employed more or less actively, in accordance with the demands of the case.

Should eclampsia occur during the course of the disease, the same general measures will be appropriate that have been advised in Chapter II. for the relief of uræmic convulsions. In addition to these, certain features of puerperal nephritis and eclampsia render a consideration of some special measures for the relief of the latter necessary.

Puerperal eclampsia presents a favorable field, in most cases, for the beneficial effects of bleeding. As a rule, these patients have not been reduced by previous disease, or by the usual loss of blood from the kidneys consequent on acute renal inflammation from other causes. I believe these facts have much to do with the cause of the greater frequency of convulsions in puerperal than in other forms of acute nephritis. If my conjectures be correct, they also explain the greater efficiency of bloodletting in convulsions in puerperal than in other forms of nephritis. However this may be, the fact remains that bloodletting has proven particularly efficacious in puerperal eclampsia; accumulated experience has so strongly confirmed this that it no longer admits of doubt.

The best results of venesection in cases of eclampsia follow the rapid withdrawal of blood. The incision should, therefore, be such as to insure a free and full stream. If the patient be robust, and the convulsions violent, from sixteen to twenty ounces of blood may be promptly taken from the arm. Should convulsions recur after the first bleeding, the vein should be reopened, and one-half to two-thirds the previous quantity of blood be withdrawn. The quantity of blood taken should be graded to correspond with the capability of the system to bear the loss. If the patient be debilitated or anæmic, it will be better to avoid venesection, and to rely upon other measures.

Recent experience has shown that the hypodermatic injection of morphine is efficient in relieving the convulsions due to puerperal nephritis; in fact, it is one of the most reliable measures for this special purpose at our command. The use of morphine in convulsions has already been considered in the chapter on uræmia. It only remains in this connection to call attention to the fact that large doses of morphine often act injuriously upon the fœtus if administered to the eclamptic patient before the termination of gestation. The smaller doses should therefore be used in these cases before delivery, unless we are actually driven to use larger doses in order to preserve the life of the patient.

Inhalations of nitrite of amyl have been employed for the purpose of controlling puerperal convulsions. On the motor centres of the cord the nitrite of amyl acts as a direct and powerful depressant, at the same time it exerts a similar but less pronounced action on the nerves and muscles, decreasing, but not destroying, their functional life (H. C. Wood). Dr. Wm. F. Jenks reports a case of puerperal convulsions occurring immediately after delivery, in which the paroxysms were instantly arrested by inhalations of nitrite of amyl. Its use, however, was followed by relaxation of the firmly contracted uterus and very alarming hemorrhage. (*Philadelphia Medical Times*, 1872, vol. ii. p. 404.)

Dr. Green, of Sandown, and others report good results to have followed the administration of nitroglycerine in cases of puerperal eclampsia.

The power of most of the nitrites in allaying reflex muscular spasms cannot be successfully denied; but their powerful vaso-dilator influence over the arterioles permits the reflex centres to be unusually exposed to the action of effete products with which the blood is surcharged in these cases. Moreover, the nitrites, by diminishing arterial tension, also lessen the filtration of urine, and they reduce to a very marked extent the oxidizing powers of the blood; all of which, we are taught, favor the production of acute uræmia. Further observations are therefore necessary in order to determine the range of safety and utility of the nitrites in these cases.

Lastly, Prof. V. G. Lashkevitch has pointed out that oxygen possesses considerable power to lower increased reflex action, and suggested that inhalations of this gas might prove of service in cases of puerperal eclampsia. Acting on this suggestion, Dr. Favr, of Kharkov (*Vratch*, No. 13, 1885), used oxygen inhalations in two cases of puerperal eclampsia with brilliant results. Although warm baths, wet packings, chloral hydrate, and chloroform, failed to give relief, inhalations of oxygen were followed in both cases by prompt cessation of convulsions, with return to consciousness and complete recovery (*London Medical Record*, June 15, 1885).

The advisability of terminating gestation by artificial means is an important question to be considered in cases of puerperal nephritis.

If nephritis be discovered during the fourth, fifth, or sixth month of gestation or even later; and if the renal complication refuse to yield to appropriate treatment, the termination of gestation will afford the most certain means of arresting the renal inflammation, and preventing permanent damage to the kidneys. The above course is doubly advisable in cases in which nephritis complicated a previous gestation.

When eclampsia appears during the last weeks of pregnancy, the practice heretofore generally advised has been to terminate gestation as quickly as possible by artificial means. It is well to bear in mind that nephritis, rather than pregnancy, is the *chief cause* of the convulsions in these cases; and, therefore, the termination of labor does not arrest the convulsions immediately, although it usually does so in time if the patient survive, because it arrests the nephritis upon which the convulsions depend. It should also be borne in mind that the period of labor is attended by greater danger of convulsions than is any other period; for statistics show that more cases of eclampsia arise during labor than before and after labor combined.

Since then artificial delivery controls convulsions chiefly through modifying the nephritis, it is proper first to consider the probabilities of being able to accomplish this by other measures before resorting to the former. We know that if it be possible to reëstablish the function of the kidneys, and to maintain the quantity of urine normal or nearly so, that the convulsions will cease. We also know that in many cases of puerperal convulsions the nephritis has either been cured, or so far modified by treatment after the occurrence of eclampsia, that the patients have subsequently passed through labor without further accident. The importance, therefore, of the vigorous employment of those measures most likely to modify the renal inflammation and restore the urinary secretion should not be lost sight of at the same time that treatment is being employed for the purpose of controlling the convulsions. This practice is the more successful since Dr. Loomis has demonstrated that puerperal convulsions can frequently be held in check by the hypodermatic injection of morphine; thus affording time to reëstablish the functions of elimination and modify the nephritis.

If, however, the convulsions be so violent and frequent that the patient's life be endangered, and other measures have failed to relieve them, it will be proper to resort to artificial delivery without further delay, as such course will then afford the best means of arresting the convulsions and preserving the life of the mother and also that of the child.

If eclampsia occur after labor at term has set in, the convulsions should be brought under control if possible by the means already described, and the labor terminated as rapidly as possible. It does not fall within my province to discuss the most desirable method of hastening labor in these cases further than to suggest that such measures should be employed as are likely to accomplish rapid delivery with the least amount of muscular exertion on the part of the patient.

Should nephritis consequent to pregnancy continue after the termination of the puerperal state, the disease and its complications will require substantially the same treatment as nephritis due to other causes.

CHAPTER VIII.

LARDACEOUS DEGENERATION OF THE KIDNEYS.

LARDACEOUS, or, as it is sometimes called, amyloid degeneration, is not a primary renal disease; but is the result of a constitutional defect, which, in most cases, is traceable to well-known causes. The kidneys are not the only, nor, in many cases, the primary seat of the anatomical changes induced by the disease; for the same changes occur simultaneously in other organs, especially in the more vascular ones, as the liver, spleen, etc.

Pure lardaceous disease in the kidney is not a form of nephritis, although it is frequently preceded or accompanied by renal inflammation. It is often incorrectly classed as a form of Bright's disease.

ETIOLOGY.

Lardaceous disease is generally believed to be due to changes in the composition of the blood, the exact nature of which is not yet clearly understood. Bartels has suggested that "perhaps some chemical agent" is formed in the process of suppuration, the absorption of which into the circulation, either by subsequent deposit in, or by mere contact with the tissues, excites the degenerative changes. Dickinson holds that the lardaceous change is connected with deficiency of the normal alkaline condition of the blood, in support of which he has shown that amyloid matter, although closely allied to fibrin in other respects, is greatly deficient in the salts of potassium. Whatever the precise chemical changes in the blood in these cases, the disease is most frequently associated with chronic suppurative processes, or cachectic conditions of the system, especially syphilis, tuberculosis, and scrofula.

Chronic Suppuration.—This is the most frequent exciting cause of lardaceous change in the kidneys. The location of the suppurative process, except, perhaps, that in bone (necrosis), does not seem to influence the production of this disease so much as the chronicity of the suppuration. Necrosis of bone, suppurating cavities in the lungs, dilated bronchi with bronchial catarrh, atonic ulceration of the skin, ulceration of the bowels, fistulæ, psoas and lumbar abscesses, ulcerating cancer, especially uterine, and, lastly, chronic nephritis, and pyo-nephrosis, are among the most common causes of the disease.

Bartels has insisted that suppuration of itself is not sufficient to bring about lardaceous change, unless the suppurating surface be exposed to the air. It may be that lardaceous change is more frequent under the latter circumstances, but that atmospheric contact with the suppurating surface is necessary in order to bring about the lardaceous change has been disproven by Dr. Belfield (*Diseases of the Urinary and Male Sexual Organs*, 1884).

Tuberculosis.—Pulmonary tuberculosis is often followed by lardaceous disease, even in cases in which the tuberculosis does not result in the formation of suppurating cavities in the lungs. Wagner found lardaceous disease in seven per cent. of his tuberculous patients.

It has been observed that the progress of pulmonary tuberculosis becomes checked in many cases upon the appearance of lardaceous disease. I am able to corroborate this statement from personal observation.

Syphilis.—From what has already been said of suppuration as a cause of lardaceous disease, it might be expected to arise from the ulcerations of the throat and genital organs, and the extensive necrosis of bone, so common in syphilis; but, aside from these, lardaceous disease sometimes springs from syphilis at a comparatively early stage of the latter, before any of the more destructive processes of ulceration have been set up. More often, lardaceous disease arises late in the course of syphilis; sometimes so late that the active symptoms have passed away, and in such cases the true cause of the disease may be overlooked.

It is well to bear in mind a fact which is of practical significance in this connection, viz., that lardaceous disease frequently occurs in syphilitics, whose general condition is well preserved. This fact favors the view that syphilis itself is the cause rather than the ulcerations or states of the system induced thereby.

It is often impossible to trace the disease directly to any of the causes mentioned; but even in such cases it is strongly probable that some constitutional taint—perhaps of hereditary origin—has given rise to the lardaceous changes without having manifested other symptoms sufficiently marked to direct attention to its presence.

MORBID ANATOMY.

In the early stage of the disease the kidneys are usually found to be larger than normal. The capsule strips off easily, leaving a perfectly smooth, pale yellow, anæmic looking surface. On section the same appearance is observed throughout the cortex. The cortex is seen to be thickened—being to the medulla as one to two, instead of one to three or four, as in the normal condition. The Malpighian bodies are seen scattered through the cortex as dull, semitranslucent spots; they are prominent because infiltrated with lardaceous matter. The medulla is not altered, except, perhaps, in the more distinct striation of the rays and abnormal paleness of the pyramids.

If an aqueous solution of iodine be poured over the cut surface, it will be observed that parts of the latter instantly turn a dark mahogany color. This is especially noted in the Malpighian bodies and vasa recta, as these are usually the first to take on the lardaceous change. The normal tissues remain unchanged, but every point where lardaceous change has set in is clearly indicated by the brown staining of the iodine solution.

Methyl-aniline violet stain is another beautiful and striking test for lardaceous matter, which it turns red; at the same time it colors the normal structures pale blue, and, therefore, presents the diseased tissues in very strong contrast.

If a thin section be examined (unstained) under a low power, the Malpighian bodies appear enlarged and glassy; though this translucency may not affect the whole tuft, but only some of its vessels, others remaining normal. The tufts are enlarged so that there is little or no space between them and the capsules of Bowman. The vessels of the tufts are seen to be much thickened, as are also the afferent arterioles and interlobular arteries.

It is desirable, for purposes of closer study, to stain sections with iodine, or methyl-aniline, and examine them under a high power (x 300.) It will then be noted that the degeneration occurs irregularly in patches, some of which are very limited. On examining the vessels—especially those of the tufts—only one side of some of the vessels is

254 LARDACEOUS DEGENERATION OF KIDNEYS.

seen to be affected, while the other side may be normal. Some of the vessels of the tufts escape completely, while others become wholly involved. The lining of Bowman's capsule usually escapes change in this stage. The muscular coats of the arterioles and vasa recta are seen to be greatly thickened, and their lumens are much narrowed.

FIG. 18.

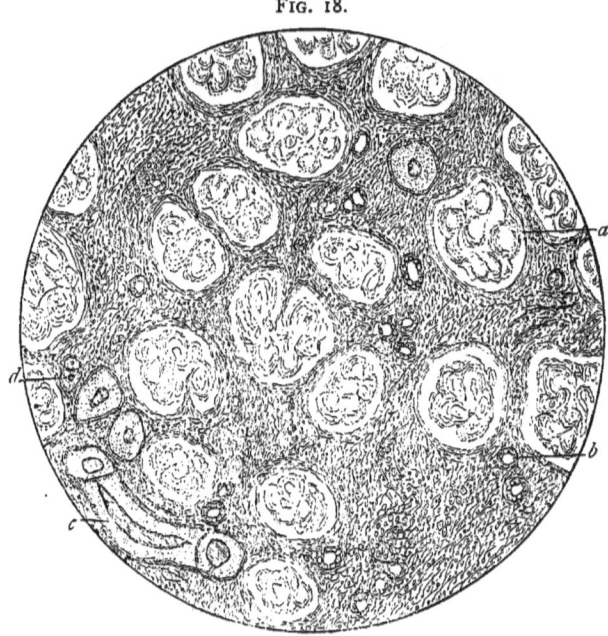

Lardaceous degeneration of kidney; cortical section. *a.* Lardaceous Malpighian tuft. *b.* Tubule compressed, seen in cross-section. *c.* and *d.* Arteries showing lardaceous change.

In the more advanced stages of the disease the organs are usually considerably enlarged, and the anatomical changes just described are more extensive. In these cases the lardaceous change includes the whole tuft of vessels in some places, and also the lining of Bowman's capsule;

even the basement membrane of the convoluted tubes may participate in the degenerative change. The epithelium of the convoluted tubes becomes fatty and breaks down, but rarely takes on the lardaceous change. The most extensive changes are always found in the small arteries, few of which escape in well-marked cases.

Prof. Greenfield gives the following order in which the various parts become the seat of lardaceous change: (1) Afferent arterioles. (2) Groups of glomerular capillaries, especially those of the superficial cortex. (3) Arteriolæ rectæ. (4) Efferent arterioles, and the capillaries into which they break up. (5) Capsule of Malpighian body. (6) The capillaries which run between the bundles of straight tubes. (7) The basement membrane of the convoluted tubules. (8) Large interlobular arteries. (9) Walls of the straight tubules, especially near the papillæ. (10) Large branches of arteries and veins in the boundary area. (11) The connective tissue around the collecting tubules at the tips of the papillæ. (12) The epithelial cells, rarely.

In many cases, especially in the early stages of lardaceous disease, the kidneys appear normal to the naked eye; but upon application of the iodine solution the lardaceous change becomes instantly visible. It is often to be observed on application of the iodine solution that lardaceous change has become engrafted upon chronic nephritis. This is the more likely to be overlooked, since the organs in both conditions are often to the naked eye very similar in appearance. It is earnestly advised, therefore, that some test solution, such as iodine, be always applied at the autopsy in those cases in which the kidneys pass under inspection.

The liver is found to be enlarged in many cases, sometimes greatly so. It is pale, anæmic, and waxy in appearance, and translucent at the edge of the section. The small arteries show the same degenerative changes as the kidneys, and respond to the iodine solution in a similar manner.

The spleen is even more frequently found enlarged than the liver. The small arteries and Malpighian bodies are the principal seats of the lardaceous change in this organ.

The small arteries of the intestines are frequently found to have undergone lardaceous change, resulting in lesions in the intestinal mucous membrane.

Less frequently, lardaceous changes are found in the vessels of the pancreas and in lymphatic glands.

Prof. Stewart, of Edinburgh, has noted lardaceous degeneration in the vessels of the heart; Dr. Bennett has seen lardaceous changes in the vessels of the placenta; Dr. Gardiner has found lardaceous change in cancerous structures; and some observers speak of having found it in the vessels of the lungs and brain.

SYMPTOMS.

The appearance of a patient who is the subject of lardaceous disease, even in its earliest stages, impresses the physician as one who is plainly out of health. These patients are anæmic, in most cases, in the earliest periods of the disease, and more or less anæmia is usually present for a considerable period of time preceding the actual appearance of the disease. In addition, the skin usually presents a sallow, cachectic hue. As a rule, muscular weakness, lassitude, and disinclination for exercise, are early accompaniments of the disease. While these general premonitory symptoms are pretty constant, there is but little in the local manifestations of the disease in the renal organs which is likely to attract attention in the early stage.

The Urine.—The quantity of the urine is subject to considerable variation in the different stages of the disease. It also frequently varies in the same stages in different cases, depending upon the presence or absence of coincident nephritis, diarrhœa, etc. These facts probably account for the different opinions expressed by various observers as to the quantity of urine passed in this disease.

It may be stated, however, as a general rule, that the quantity of urine in cases of lardaceous kidney is increased, even in the early course of the disease. In many cases this increase is very marked; in most, if not in all cases, the quantity exceeds the normal, sooner or later. The very marked increase in the quantity of urine sometimes observed, occurs chiefly during the height of the malady. In the last stage of the disease the urine suffers considerable reduction in quantity; but it is not usually reduced below 500 c.c.

The specific gravity of the urine corresponds inversely with its quantity. In some cases it sinks very low—I have seen it sink to 1.005. More frequently, especially during the height of the disease, the specific gravity ranges from 1.005 to 1.015. If lardaceous disease of the kidneys becomes complicated by nephritis, the specific gravity of the urine may rise above normal—sometimes reaching 1.030 or over. The urine is usually remarkably clear, and its reaction slightly acid in uncomplicated lardaceous disease.

The percentage of urea and solids varies considerably, owing doubtless, in some measure, to the variation in the quantity of the urine. As a rule, the urea and solids are below normal throughout, though they do not suffer so great a reduction in this disease as in most forms of nephritis.

In uncomplicated forms of lardaceous disease but little sediment is found in the urine. If carefully collected and placed under the microscope this deposit will be found to contain a few clear hyaline casts, lymph-corpuscles, and some epithelial cells from the convoluted tubes, some of which may be partly fatty. Occasionally waxy looking, highly refracting casts may be seen; and, in late stages of the disease, they are said occasionally to give the characteristic reaction of lardaceous matter with iodine solution.

The urine contains albumin in lardaceous disease of the kidneys. Exceptions are extremely rare, and confined chiefly to the earliest stage of the disease, before changes in the vascular tunics become pronounced.

The percentage of albumin in the urine is usually large, even from a very early period. It often reaches two or even three per cent. by actual weight. As a rule, the quantity of albumin excreted by lardaceous kidneys continues to be quite large, although the actual percentage fluctuates with the quantity of urine passed. Senator, Edlefsen, and others have noted that the urine in cases of lardaceous disease often contains paraglobulin, which may exceed in quantity serum albumin. In some cases an excess of indican has been observed.

The chlorides in the urine are greatly reduced; the phosphates also undergo marked reduction.

Dropsy.—The presence of dropsy in lardaceous disease is very inconstant. The slighter forms, such as œdema of the feet and ankles, may be present quite early; while, again, dropsy may remain absent throughout the whole course of the disease. These facts tend to the conclusion that the dropsy of lardaceous disease is not dependent upon the renal complication; but upon some other cause, such as hydræmia, which is often marked in these cases.

Roberts has suggested that the dropsy in lardaceous disease is due chiefly to obstruction of the portal circulation consequent to lardaceous changes in the liver and lymphatic glands surrounding the portal vein.

Probably in the majority of cases dropsy becomes associated with lardaceous disease sooner or later. In most cases it is confined to the lower extremities and abdominal cavity (ascites). If dropsy once becomes established in these cases, it is more difficult to control, than when caused by nephritis. I once succeeded in reducing completely a marked dropsical condition dependent upon lardaceous disease, by means of a course of hot-air baths. The dropsy quickly returned, however, when the baths were discontinued, and it continued to the end.

The Digestive System.—The consequences of lardaceous disease fall with special force and frequency upon the digestive system. It is not uncommon to observe the effects of lardaceous disease in the intestinal tract before any evidences of the disease are manifested by the kidneys.

The appetite is impaired from the beginning in the great majority of cases; indeed, anorexia is often present for a long time previous to the appearance of albumin in the urine. Nausea and vomiting often occur when the disease becomes established. These symptoms sometimes prove very obstinate. The matter ejected by the stomach is usually acid in reaction, and has a watery appearance.

In a limited number of cases—especially in those in which the disease has a syphilitic origin—the stomach remains undisturbed and the appetite good; at least in the early course of the disease. But from whatever cause the disease may arise, few cases reach an advanced stage without more or less disorder of the stomach being present.

Disorder of the bowels is one of the most constant accompaniments of this disease. In the early stages constipation is frequently present; later, mild attacks of diarrhœa are common. When the disease becomes well established, diarrhœal attacks become very frequent, seriously augmenting the patient's debility. Finally, in the late stages of the disease, the diarrhœa becomes extremely obstinate, owing to the presence of extensive lesions in the intestinal vessels, and consequent ulcerations of the intestinal mucous membrane. Treatment now rarely affords much relief from this complication, and the continued diarrhœa soon leads to exhaustion and death of the patient.

The Circulatory System.—Although lardaceous disease always involves extensively the smaller bloodvessels, yet, with the exception of anæmia, the symptoms referable to the circulatory system are rarely notable. Anæmia is almost invariably present from the earliest period; indeed, in the great majority of cases it precedes the appearance of the disease in the renal organs. In the early stage of the disease anæmia is not so extreme as in the early stage of nephritis, but as the disease continues it becomes more and more marked.

In consequence of the changes in the small arteries, the more vascular organs become greatly altered. The spleen presents the most frequent changes, in most cases becoming greatly enlarged. Cardiac changes are rare in uncomplicated lardaceous disease. Cardiac hypertrophy is seldom met with, even in the late stages of the disease; probably only when cirrhosis or chronic nephritis becomes associated therewith. In some cases the hemorrhagic diathesis, so-called, is developed, resulting in troublesome hemorrhage. Thrombosis, especially of the femoral veins, is not uncommon in this disease.

The Glandular System.—The most striking change in this connection is the frequent increase of the area of dulness over the liver, consequent on enlargement of that organ. While enlargement of the liver is less frequent than enlargement of the spleen, yet it occurs in more than fifty per cent. of these cases. Sometimes the liver attains an enormous size. In consequence of the degenerative changes which usually occur in the liver, the function of that organ frequently becomes disturbed, the most apparent evidence of which is the semijaundiced hue of the skin so often observed in these patients.

The pancreas and abdominal lymphatic glands frequently share in the degenerative changes common to lardaceous disease, and it is likely that the extreme malassimilation occurring in many cases is largely due to these changes.

The Nervous System.—The symptoms referable to the nervous system are chiefly of a negative character. Uræmia rarely occurs, save in the late stage of the disease, when the renal organs have become the seat of extensive anatomical changes. The rarity of uræmia is chiefly due to two circumstances: first, the general condition of these patients is reduced before the renal disease sets in, and, consequently, tissue changes are not active; second, the renal epithelium is but slightly affected, and is, therefore, able to maintain its function of elimination of the solids in proportion to the demands made upon it by the reduced requirements of the system. If the patient survive until the changes in the kidneys become extreme, the function of these organs usually becomes impaired, and uræmia may result. Even under such circumstances the more active uræmic symptoms—especially convulsions—are rare. I have seen death occur in such cases from uræmic coma which had approached gradually; but I have never seen uræmic convulsions in uncomplicated lardaceous disease.

The Respiratory System.—Except in those cases in which pulmonary tuberculosis is coincident with or precedes lardaceous disease, disorders of the respiratory system are not common in this disease. Even in such cases, as has been previously remarked, the pulmonary disease often ceases to extend when lardaceous disease becomes developed.

Dyspnœa from so-called renal asthma, or from œdema of the lungs, is almost unknown in simple lardaceous disease of the kidneys.

If the patient survive till the lesions in the kidneys have become very extensive, or if nephritis become superadded to lardaceous disease, pneumonia and pleurisy may occur.

The secondary inflammatory complications which are so common in nephritis, are comparatively rare in simple lardaceous disease of the kidneys. Peritonitis is the one most frequently met with; it is usually associated with purulent effusion into the peritoneal cavity, and, as a rule, quickly proves fatal.

COURSE AND DURATION—The course of lardaceous disease is subject to considerable variation in individual cases, depending upon the cause and nature of the complications which arise. It is essentially a chronic disease, and, unless the patient succumb to some complication or morbid constitutional condition which caused the disease, he will usually survive for several years.

The disease begins in the most insidious manner, and it may continue for several months before it gives rise to symptoms sufficiently pronounced to direct attention to its presence. Anæmia and increasing debility are perhaps the only manifest indications of the disease in its earliest stages, and even these are frequently absent in cases of syphilitic origin. If the urine be examined, it will be found to contain albumin and usually a few pale casts; but there is so little apparent disturbance of the urinary function in these cases for a long time, that a chemical examination of the urine is usually delayed until suspicions of the disease are aroused by more advanced symptoms.

As the disease continues and becomes well established, it is usually associated with more marked and characteristic symptoms. There is more or less increase in the quantity of urine—the patient sometimes rising at night to urinate. Diarrhœal attacks are frequent, and the appetite is more or less impaired. Albuminuria is now more pronounced, as are also debility and anæmia. Œdema of the feet and limbs may or may not be present, but more or less enlargement of the liver and spleen is usually to be found. The disease may continue with moderate fluctuation of above symptoms for from a few months to several years.

When the disease reaches an advanced stage, diarrhœa becomes more constant and more severe. Vomiting is likely to become a distressing and obstinate symptom. The liver often becomes enormously enlarged in this stage of the disease, and is usually tender, so that the patient may be unable to lie on his right side. Dropsy is usually associated with the more extreme grades of hepatic enlargement. The urine is now highly albuminous, and contains broad waxy casts, as well as granular and fatty casts. Anæmia and debility become very prominent features of this stage of the disease. Finally, the urine diminishes in quantity and some secondary inflammation —most often peritonitis—may set in and prove fatal.

The course of the disease is usually most rapid in cases which arise from chronic suppuration. If the disease arise from phthisis, especially early phthisis, its course may be more tardy. Cases arising from syphilis often pursue an extremely chronic course, and moreover, in such cases, improvement, and even cure, may be brought about after the disease has lasted five years or longer.

These patients more often succumb to the general exhaustion consequent upon diarrhœa or prolonged suppuration, than to the immediate consequences of the changes in the renal organs.

The duration of the disease is extremely variable. In some cases death takes place in a few months after the discovery of the disease, more frequently the disease lasts one or two years or more; and, in exceptional cases, it has been known to continue from seven to ten years before reaching a fatal termination.

DIAGNOSIS.

CONSTRUCTIVE.—In well-marked cases of lardaceous disease of the kidneys but few difficulties will be encountered in recognizing the disease.

The urine is usually highly albuminous, pale in color, of low specific gravity, increased in quantity, and contains but little sediment, in which may usually be found a few hyaline casts. The general condition of the patient is debilitated; and a history of tuberculosis, syphilis, or chronic suppuration may usually be made out. Dyspepsia and diarrhœa are seldom absent for any considerable period of time. Dropsy may be present, and if so, is confined chiefly to the lower extremities and the abdominal cavity. If, in addition to the above symptoms, distinct enlargement of the spleen or liver be recognized, there can be no further doubt that the case is one of lardaceous disease.

In some cases, however, especially in the early stage of the disease before the spleen becomes enlarged, it may be difficult to recognize the disease; but even in these cases certain symptoms are likely to be present which should lead us strongly to suspect the presence of lardaceous disease. These are debility and cachexia, diarrhœa, anæmia, evidences of tuberculosis or syphilis, or the existence of chronic suppuration, coupled with albuminuria; the urine at the same time being of low specific gravity, of light color, increased in quantity, nearly free from sediment, and containing a few hyaline casts.

DIFFERENTIAL.—The diseases most likely to be confounded with lardaceous degeneration of the kidneys are chronic nephritis and cirrhosis of the kidneys. The means of distinguishing these affections from lardaceous disease have been fully considered in Chapters IV. and V.

PROGNOSIS.

Generally speaking, the prognosis in lardaceous disease is unfavorable. It is true that recovery sometimes follows even after the disease has attained considerable headway, but the majority of cases terminate fatally.

The prognosis in individual cases will depend largely upon the causes which bring it about. The most hopeful cases are usually those which are the outgrowth of syphilitic conditions, as the latter are ordinarily amenable to treatment. In some cases the primary disorder is incurable, especially if advanced, and in these cases the prognosis is necessarily unfavorable.

The prognosis is more favorable in children than adults, possibly because the recuperative powers of the system are much greater in the former than in the latter.

In those cases in which nephritis becomes superadded,

the prognosis must always be considered unfavorable. While the ultimate result of lardaceous disease is generally fatal, yet the immediate outlook is often encouraging; for, in many cases, life may be greatly prolonged by appropriate treatment. Indeed, we are seldom justified in withholding all hopes of recovery, as cases have frequently occurred in which complete recovery took place after the disease had attained a marked state of development, including pronounced enlargement of the spleen and liver.

TREATMENT.

The general treatment of lardaceous disease will depend largely upon the nature of the primary disease from which it springs. As this may be of a syphilitic, tuberculous, or strumous character, the range of measures applicable in all cases is very wide; and a systematic consideration of them all would not fall within the scope of this work. Therefore, there will be considered under this head only those measures which are conceded to be of special utility in lardaceous disease of the kidneys when it springs from these states.

With regard to syphilis, it may first be stated that fewer cases of lardaceous disease would arise from this cause if specific treatment were continued, as it should be, for two or three years after the appearance of the primary lesion. Too often, syphilis is merely lulled into temporary quiescence by the employment of specific medication for a few months. Among the many sad consequences of this incomplete course of treatment for syphilis is the not uncommon one of lardaceous disease. In all cases of lardaceous disease of the kidneys, therefore, careful inquiry should be made, not only for evidences of syphilitic infection, but also as to the length of time the patient has been subjected to specific treatment.

If we have to deal with a case of lardaceous disease of the kidneys consequent upon neglected syphilis the iodide of potassium will usually prove of the greatest service in its treatment. It should be administered in as large doses as the stomach will tolerate—from ten to thirty grains three times a day. If the use of this agent brings the disease under control —which happily it sometimes does—we have next to consider the propriety of administering mercury with a view to eradicate the syphilis. In this matter we should be governed by the nature and duration of the specific treatment which the patient may have previously received and the stage of syphilis in which the disease arises.

In cases in which lardaceous disease is the result of tuberculous or strumous conditions, appropriate measures for the relief of these are likely also to prove the most efficient in controlling the renal complication. In addition to the agents commonly employed, iodoform and chloride of calcium have been suggested as especially applicable in the treatment of lardaceous disease due to the above conditions. Iodoform may be administered in doses ranging from one to five grains, in the form of pill, repeated two or three times a day. It should be borne in mind that if it be continued long, especially in large doses, it is liable to induce very serious disturbance of the system known as "iodoform poisoning."

The chloride of calcium is especially applicable in cases associated with enlargement of the lymphatic glands. The dose for adults is from five to twenty grains, repeated two or three times a day; it is best given largely diluted with milk.

In cases depending upon necrosis of bone, or chronic

suppurative processes, antiseptic surgery—for which we owe so much to Sir Joseph Lister—is often the most efficient means at our command for the purpose of favorably influencing this disease. With its aid we are often able to control the suppurative processes which originated the disease; furthermore, it enables us to undertake successfully the removal of a diseased limb, or part, which may be keeping alive the disease. It is a fact of practical significance in connection with this that in hospitals where the antiseptic system of treating suppurative diseases has been adopted, the occurrence of lardaceous disease has been less frequent.

The adynamic condition of these patients calls for a liberal and sustaining diet. Fortunately, the functional capacity of the kidneys is usually such that a liberal diet may be permitted without incurring danger from uræmia. The most nutritious diet compatible with the digestive powers of the patient should, therefore, be used. Every means should be employed which is likely to improve the digestive and assimilative powers of these patients, which are almost invariably impaired. Moderate exercise in the open air should be insisted upon when the weather is fine, provided that the patient be able to go about. If the strength of the patient do not permit of walking, he should be induced to ride out as often as circumstances will allow.

Dyspepsia is best met, in addition to the above measures, by such remedies as strychnia, mineral acids, pepsin —especially wine of pepsin—small doses of quinine, and such vegetable bitters as gentian, quassia, etc. A good wine—especially Burgundy—often proves highly beneficial to these patients when served with the principal meals.

Of all the symptoms of lardaceous disease, diarrhœa is usually the most troublesome one with which we are called upon to contend. If we have reasons to believe that the diarrhœa is due to simple intestinal catarrh, it should be treated by the administration of mild astringents and small doses of opium. Small doses of acetate of lead are very effectual in most of these cases, though its use should not be continued over an extended length of time, as in such case it sometimes acts as a renal irritant. If there be coincident nephritis, lead should not be administered. Dr. Ralfe (op. cit.) speaks highly of the use of bismuth in these cases.

On account of the atonic condition of the stomach in these cases, the opiate should be administered in such form as is least likely to disturb that organ. For this purpose, the deodorized tincture of opium is the most suitable, and it is still less likely to disorder the stomach if given per rectum.

In thóse cases in which diarrhœa is due to intestinal ulceration the alterative astringents are the most reliable agents to combine with the opiates for the purpose of controlling the diarrhœa. The use of nitrate of silver is often followed by good results in these cases. The sulphate of copper, however, outranks all other drugs in point of efficiency in controlling this special form of diarrhœa. It should be administered in doses of one-eighth to one-fourth of a grain in the form of pill, and it may be repeated every four to six hours. The sulphate of copper may often be advantageously combined with extract of belladonna in doses of one-eighth to one-fourth of a grain which may be given by the mouth, while the opiate is best administered by the rectum.

Dropsy sometimes becomes sufficiently marked to call for special treatment, especially in the late stage of the disease. The most appropriate agents to employ in such cases are diuretics and hot air baths; the use of cathartics should be avoided, especially the more active ones.

In cases in which no special cause can be discovered to account for the development of lardaceous disease, it will be necessary to resort to the use of certain remedies which have been found more or less efficient in controlling the disease. While, unfortunately, as yet we possess no means upon which we can rely for the purpose of arresting lardaceous disease, yet certain drugs would seem to exercise a favorable influence over the disease in a limited number of cases. Of these the iodides are probably the most efficient. The iodide of iron is the most appropriate preparation for administration in most of these special cases. This may be given in the form of syrup, or, better still, Blancard's pills—the latter being an especially reliable preparation. Dr. Ralfe (op. cit.) believes that the efficacy of the iodides in these cases is greatly increased by the addition of arsenic.

Bartels relies on the iodide of potassium as the only means of directly influencing the degenerative changes for the better in all cases.

The administration of iron is followed by good results in most cases. Iron does not agree with all anæmic patients, as, for instance, in some cases of phthisis; and this is true, though to a less extent, in lardaceous disease.

In most of these cases iron should be given with a view to modify the anæmic condition of the patient; but if it prove unsuitable, after a fair trial, its further use should be discontinued.

One of the most serious and frequent complications of lardaceous disease of the kidney is nephritis. If this arise during the course of the disease, its treatment, including that of its complications, should be conducted substantially upon the same principles advised for the treatment of nephritis in Chapters III. and IV.

CHAPTER IX.

CYANOTIC INDURATION OF THE KIDNEY.

THIS is a secondary renal disease, depending upon obstructions to the circulation which retard the flow of blood through the kidneys. It is, in fact, a condition of passive or venous congestion of the kidneys.

Cyanotic induration of the kidneys was for some time classed as a form of Bright's disease, which was improper, for, in all cases, it is the result of some other disease.

ETIOLOGY.

Cyanotic induration of the kidney is most frequently met with as the result of those diseases of the heart which obstruct the venous circulation, as mitral insufficiency, or stenosis of the mitral or aortic orifice. Less frequently, the disease is met with as the result of certain pulmonary diseases which interfere with the circulation, such as emphysema, chronic pneumonia, pleurisy, etc.

It will be remembered that the abdominal veins, including the renal, are unprovided with valves, and, therefore, when cardiac or pulmonary obstruction to the circulation arises, the venous blood is dammed back through the inferior vena cava and renal veins into the kidneys, resulting in venous engorgement in those organs.

Phthisis rarely gives rise to cyanotic induration of the kidney. The diminished vascular capacity of the lungs in phthisis is accompanied by pronounced anæmia, and the comparatively small quantity of blood in the system finds its way through the remaining pulmonary vessels with ease.

Local obstructions to the return of the venous blood from the kidneys are not often met with. Thrombosis of the inferior vena cava above the junction with the renal veins has been known to give rise to cyanotic induration of the kidneys. The inferior vena cava may become compressed by large growths immediately in front of it, and this may result in more or less stasis of venous blood in the kidneys, but cyanotic induration of the kidneys is very rare from this cause. As Bartels has remarked, "only the higher grades of general venous stasis exert any appreciable effect upon the kidneys and their functions."

Practically, therefore, the condition of the kidneys under consideration is chiefly met with in obstructive cardiac disease, especially valvular lesions at the mitral orifice; for these only produce the more intense degrees of general venous stasis.

MORBID ANATOMY.

The kidneys are usually found enlarged in the early stages; while, if the disease has continued very long, the organs may be found slightly smaller than normal.

The surface of the kidney is congested, and of a purple color, and in very chronic cases it may present a granular appearance. The capsule is usually non-adherent, and on its removal the venæ stellatæ are seen to be very prominent. The kidney is hard, tough, and elastic, feeling not unlike India rubber to the fingers.

On making a section through the organ the cut surfaces turn a deep crimson on exposure to the atmosphere. The cortex is seen to be slightly thickened, except in very chronic cases. The Malpighian bodies are usually enlarged, and the interlobular veins are prominent and distended.

The medulla is seen to be the most deeply congested; its cut surface being of a deep purple color. The venæ rectæ are especially distended and congested, giving to the pyramids a striated appearance by reason of their contrast with the interlying bundles of straight tubes.

Under a high power the prominence of the Malpighian body is still more apparent; the tuft is seen to fill, and, in some cases, to distend the capsule of Bowman. Some of the capillaries of the tufts are ruptured, as extravasations of blood are sometimes seen within the capsules of Bowman, and even in the convoluted tubes. The epithelium of the convoluted tubes is usually but slightly altered. It may be swollen, and sometimes granular, but it rarely becomes fatty.

In chronic cases small wedge-shaped patches of dense tissue may sometimes be seen beneath the capsule of the organ, which run inward along the course of the interlobular arteries. Within these patches the Malpighian bodies and tubules may be seen to be more or less atrophied, while the widened spaces between them are occupied by an increased amount of connective tissue.

Collections of altered blood corpuscles may be seen in the convoluted and straight tubes; and occasionally perfectly black casts ("melanin") are seen in a few of the straight tubes near the papillary margin of the medulla.

SYMPTOMS.

The general symptoms associated with cyanotic induration of the kidneys are chiefly those due to cardiac disease. Since the renal disorder is purely secondary to the cardiac changes, general symptoms of a pronounced character are usually present for some time before the renal disorder becomes established.

The Circulatory System.—The pulse is small, thready, and abnormally frequent. The veins generally are overfilled at the expense of the arteries, causing the skin to present the dusky purple hue characteristic of venous obstruction. The larger veins are prominent, especially those over the abdomen and lower extremities.

Cardiac changes are usually observable in the form of valvular disease, stenosis of the orifices, or fatty degeneration or dilatation of the cardiac walls. These are attended by the characteristic symptoms of these lesions, which need not be described here.

The Respiratory System.—Dyspnœa is nearly always present, and it is usually one of the first symptoms observed. Frequently these patients consult their medical attendant for the relief of dyspnœa, when their true condition is first discovered. The dyspnœa is much aggravated by exercise—indeed, the latter cannot be continued without causing the patient to "be out of breath."

In some cases dyspnœa is not marked at first, and it may be complained of only during active exercise. As the disease becomes advanced, dyspnœa is usually more constant, as well as more urgent, so that the patient is often unable to lie down without greatly embarrassing his breathing.

Dropsy.—This is rarely absent. It usually begins in the feet and extends to the legs, and it may involve the abdominal cavity; but the cavity of the chest and the upper extremities usually escape. The dropsy may become very marked in the limbs, and ascites may also attain a high grade, so that it may become necessary to resort to measures for its relief.

The Urine.—The daily quantity of urine usually falls below normal in this disease. The specific gravity is increased, ranging from 1.025 to 1.035 or above. The color of the urine is dark brownish-red, and the reaction is usually highly acid. The quantity of urates, especially uric acid, is relatively and sometimes absolutely increased. Free uric acid crystals are usually seen in large numbers in the sediment. As a rule, the urine contains a small percentage of albumin—perhaps as much as two-tenths of one per cent. The sediment usually contains a few small hyaline casts, and occasionally a few casts may be found with scattering blood disks attached to them. Sometimes no casts can be found in the urine.

Uræmic complications are not met with in this disease unless nephritis becomes superadded. This is chiefly due to the fact that in cyanotic induration of the kidney the renal epithelium is but slightly affected; and, therefore, the solids of the urine continue to be excreted in fair quantity.

In cyanotic induration of the kidneys, as has already been shown, the renal organs are in a constant state of venous congestion, rendering the patient unusually liable to attacks of nephritis. It is not uncommon, therefore, to observe attacks of nephritis in these patients from very slight causes.

COURSE AND DURATION.—In the early course of the disease the symptoms may be slight. Some dyspnœa is usually present, especially upon exertion; otherwise there may be but little to attract special attention.

The general condition of the patient is usually good, and digestion and assimilation may remain unimpaired till quite late in the disease.

Sooner or later, however, œdema appears in the feet and legs; and coincidently with this the urine will be found to be reduced in quantity, and usually to contain albumin and casts.

The further progress of the symptoms will depend chiefly upon the nature of the original disease. If this consists of lesions of the mitral valve, the symptoms may improve after a time as the mitral defect becomes compensated by hypertrophy of the right ventricle. In such case the urine increases in quantity and œdema of the extremities may pass away, and even albumin and casts may disappear from the urine. So long as the compensatory enlargement of the ventricle is adequate to overcome the obstruction, and to maintain a proper balance between the arterial and venous circulation, the case will continue in this favorable condition without the appearance of urgent or marked symptoms. It is not uncommon for this improved state to continue for several months, and in some cases for two or three years with but slight fluctuations.

Sooner or later, however, the compensatory hypertrophy of the ventricle passes into fatty degeneration or dilatation; and the cardiac power becomes permanently inadequate to overcome the obstruction to the circulation. The urine now decreases in quantity, and œdema again sets in. Any improvement in the condition of the patient from this time on can be only temporary. Dropsy, especially ascites, becomes more and more marked, often reaching an extreme degree. The urine diminishes in quantity progressively as the disease continues, till in the last stage it becomes very scanty. Dyspnœa harasses the patient so constantly that he gradually becomes enfeebled in consequence of the almost continuous respiratory efforts.

These symptoms continuing and gradually intensifying, the patient at length reaches a state of extreme exhaus-

DIAGNOSIS.

tion; and he may succumb to extreme dropsy, pulmonary apoplexy, or failure of the heart's action.

The duration of the disease is uncertain in individual cases. In some cases it reaches a fatal termination within a few months, while in others it may continue for several years before the patient succumbs.

DIAGNOSIS.

CONSTRUCTIVE.—Ordinarily, the recognition of cyanotic induration of the kidneys presents but few difficulties. The general cyanotic appearance of the patient, associated with dyspnœa and cardiac disease, should arouse suspicion of its presence, and lead to an examination of the urine. If, in addition to these symptoms, the urine be of high specific gravity and diminished in quantity, and if it contain a small percentage of albumin, a few pale casts, and an excess of urates, there can remain no doubt of the presence of venous congestion of the kidneys.

DIFFERENTIAL.—Cyanotic induration of the kidneys may be mistaken for renal cirrhosis. The chief distinguishing points to be kept in view are as follows: In renal cirrhosis cyanosis of the skin is absent, and œdema and dyspnœa are rarely present till the last stage of the disease is reached. The urine in cases of renal cirrhosis is usually above normal in quantity, of low specific gravity, and pale in color; conditions which are directly opposite to those in cases of cyanotic induration of the kidney.

The means of distinguishing cyanotic induration of the kidney from acute and chronic nephritis have been considered in Chapters III. and IV.

PROGNOSIS.

Cyanotic induration of the kidney does not cause death. Unfortunately, however, the diseases upon which it depends are essentially incurable, and, sooner or later, they terminate fatally.

In those cases in which the renal difficulty arises in consequence of valvular lesions of the heart, or weakened cardiac muscle, the employment of those measures which strengthen cardiac action often cause both the renal embarrassment and the dropsy to disappear for a considerable length of time. While, therefore, the conditions which bring about this disease are often susceptible of great improvement under appropriate treatment, yet the ultimate prognosis is always unfavorable, as the defects in the circulation are of such a nature as to render them irreparable.

TREATMENT.

The treatment of cyanotic induration of the kidneys should be chiefly directed toward the relief of the primary condition on which the renal disorder depends. The enfeebled action of the heart requires that the least possible demands be made upon that organ; and, therefore, rest should always be insisted upon. The condition of these patients will often materially improve by simply confining them to bed for a few weeks, thereby relieving the heart from all unnecessary work.

The patient should be warmly clad, and should occupy comfortable apartments, as artificial heat favors a more easy circulation of the blood.

The diet should be generous and nutritious, in order to sustain the weakened tone of the cardiac muscle. Digitalis has, heretofore, been the drug most relied upon for the purpose of increasing the power of the enfeebled heart. It should be administered in reasonably full doses if the stomach will tolerate it—fifteen drops of the tincture, or one or two grains of the powdered leaf, repeated every six hours. Under the use of digitalis the patient often improves greatly, the contractions of the heart become more slow and more powerful—permitting more entire filling of the ventricle, the dyspnœa and cyanosis become modified, the quantity of the urine is increased, œdema subsides, and the patient is relieved of most of the urgent symptoms.

Convallaria majallis and adonis vernalis have recently been brought forward as cardiac stimulants, and they may be employed with advantage in these cases. They are said not to be cumulative in their effects, and are therefore useful substitutes for digitalis when it becomes desirable to discontinue temporarily the use of the latter. Digitalis often acts more efficiently when administered in conjunction with adonis vernalis. The dose of the latter is from five to twenty drops in the form of fluid extract.

But the most promising drug of this class for the purpose of relieving the cardiac conditions upon which the disease usually depends is strophanthus. It will be remembered that digitalis not only increases the force of the cardiac contractions, but that it also contracts the arterioles; therefore, while it directly increases the power of the heart, it also impedes the arterial circulation and throws more work upon the heart.

According to the extended observations of Dr. Fraser, strophanthus acts upon the heart similarly to digitalis, though more rapidly and more powerfully, while it has no action on the arterioles. The advantages of strophanthus over digitalis in certain cardiac obstructions to the circulation are, therefore, very apparent, and likely to prove of great practical importance in the treatment of this disease. Strophanthus may be administered in the form of tincture in doses of two to six drops.

The condition of the kidneys does not call for special treatment unless nephritis becomes superadded; and in such cases the treatment applicable to nephritis should be resorted to.

Dropsy frequently becomes one of the most troublesome symptoms to manage. If the use of the agents already advised do not increase the flow of urine, but little may be expected from the use of diuretics. The citrate or acetate of potassium may, however, be employed, as in some cases they seem to increase the quantity of urine; they also reduce the excessive acidity of the urine commonly present in this disease, and tend to prevent consequent irritation.

Hot-air baths are sometimes useful measures for the purpose of reducing the dropsy. Dry hot air is the most efficacious, and the least objectionable form of hot bath in these cases.

Purgatives should be used with caution, more especially drastic ones, as they tend to weaken the patient. If it be determined, however, to resort to the use of purgatives for the purpose of reducing dropsy, the concentrated saline cathartics will be found the least objectionable agents of this class to employ.

If ascites attain an extreme grade, the excess of serous fluid may be withdrawn by means of a canula or aspirator, to the great relief of the patient.

BIBLIOGRAPHY.

BARTELS. Diseases of the Kidney. Ziemssen's Cyclopedia. English translation, vol. xv.
BASHAM. On Dropsy. Third edition, London, 1886.
BEALE. Urinary and Renal Derangements. 1885.
BELFIELD. Diseases of the Urinary and Male Sexual Organs. 1884.
BIRD. Urinary Deposits. Fifth edition.
BLACK. Lectures on Bright's Disease. London, 1875.
BRAUN. Uræmic Convulsions. English translation, 1858.
BRIGHT. Reports on Medical Cases. London, 1827.
BYFORD. Puerperal Convulsions. Article in Theory and Practice of Obstetrics. 1873.
CAPITAN. Recherches expérimentales et cliniques sur les Albuminuries Transitories. 1883.
CAZEAUX and TARNIER. Eclampsia. Article in Obstetrics and Diseases of Pregnancy and Parturition. 1884.
CHATEAUBOURG. Recherches sur l'albuminurie physiologique. 1883.
DICKINSON. Diseases of the Kidneys. Vol. i., Diabetes. Vol. ii., Albuminuria. Vol. iii., Miscellaneous Renal and Urinary Affections.
DONKIN. Skim-milk Treatment of Diabetes and Bright's Disease. London, 1871.
EBSTEIN. Diseases of the Kidney. Ziemssen's Cyclopedia. English translation, 1877. Vol. xv.
FOWLER. Chemical and Microscopical Analysis of the Urine in Health and Disease. New York.
GOODFELLOW. Diseases of the Kidney and Dropsy. London, 1861.
GOODHART. Structural Changes and Clinical Symptoms of Chronic Renal Disease Associated with Dropsy.
——Note of Two Cases of General Anasarca in Children without Albuminuria. Guy's Hosp. Rep., vol. xlii.
——Acute Dilatation of the Heart as a Cause of Death in Scarlatinal Dropsy. Guy's Hosp. Rep., vol. xxiv.
——Scarlatina. Article in A Guide to Diseases of Children. London. 1885.

BIBLIOGRAPHY.

GOWERS. Albuminuric Retinitis. Article in Medical Ophthalmoscopy. Second edition, 1882.
HARLEY. On Albuminuria. London, 1886.
HENOCH. Inflammatory Affections of the Kidneys. Article in Diseases of Children. English translation, 1882.
JOHNSON. Diseases of the Kidneys. 1852.
——Lectures on Bright's Disease. London, 1873.
KLEIN. Glomerulo-Nephritis. Pathological Society Transactions, vol. xxviii.
LOOMIS. Diseases of the Respiratory Organs, Heart, and Kidneys. Third edition, New York, 1882.
MAHOMED. Some Clinical Aspects of Chronic Bright's Disease. Guy's Hosp. Rep., vol. xxiv., 1879.
——Chronic Bright's Disease without Albuminuria. Guy's Hosp. Rep., vol. xxv., 1881.
——On Albuminuria and the Symptoms that Indicate its Gravity. Guy's Hosp. Rep., vol. xlii., 1883 and 1884.
MCBRIDE. The Early Diagnosis of Bright's Disease. New York, 1882.
MEIGS and PEPPER. Scarlatina. Article in Practical Treatise on Diseases of Children. Seventh edition, Philadelphia, 1883.
MILLARD. Bright's Disease of the Kidneys. Second edition, New York, 1886.
NEUBAUER and VOGEL. Urinary Analysis. English translation, seventh edition, 1879.
OLIVER. Bedside Urine Testing. Third edition, London, 1885.
PARKES. The Urine in Health and Disease. London, 1860.
PAVY. Croonian Lectures on Diabetes. 1878.
PEYER. Atlas of Chemical Microscopy. English translation, second edition, 1885.
PROUT. Stomach and Renal Diseases. Fifth edition, London, 1848.
RALFE. Diseases of the Kidneys. London, 1885.
——Clinical Chemistry. 1883.
——Diseases of the Kidneys. Year Book of Treatment. 1884 and 1885.
RAYER. Traité des Maladies des Reins. Paris, 1839.
REES. On the Nature and Treatment of Diseases of the Kidneys. London, 1850.
RICHARDSON. Uræmic Coma. 1862.
RICHARDSON, W. L. Treatment of Acute Parenchymatous Nephritis of Pregnancy. American Gynecological Transactions, 1879.
ROBERTS. Urinary and Renal Diseases. Fourth edition, London, 1885.
ROSENSTEIN. Classification of Bright's Disease. Transactions International Medical Congress. 1881.
SENATOR. Albuminuria in Health and Disease. New Sydenham Society translation, 1884.
SMITH. Scarlatina. Article in Practical Treatise on Diseases of Children. 1884.

STEWART. Bright's Disease of the Kidneys. Second edition, 1871.
STOKVIS. Recherches experimentales sur les conditions pathogeniques de l'Albuminurie. 1867.
THOMAS. Scarlatina. Article in Ziemssen's Cyclopedia. Vol. ii.
TODD. Clinical Lectures on Certain Diseases of the Urinary Organs. London, 1857.
TYSON. Practical Examination of Urine. Fifth edition, 1886.
———Bright's Disease and Diabetes. 1881.
———Causal Lesions of Puerperal Convulsions. 1879.
ULTZMAN. Pyuria. English translation, 1884.

INDEX.

ACETATE of potassium in treatment of puerperal nephritis, 244
    of lead in diarrhœa due to lardaceous disease, 269
Acute hypertrophy of heart in scarlatinal nephritis, 206
    nephritis, 86
Adonis vernalis in treatment of cirrhotic kidney, 186
    of cyanotic kidney, 279
Age as a predisposing cause of cirrhosis of kidneys, 140
    of scarlatinal nephritis, 194
Albumin, tests for, in urine, 34
    quantitative, 47
Albuminuria, 17
    extrarenal or false, 18
    renal or true, 22
        causes of, 24
        significance of, 31
        hygienic treatment of, 49
    in cirrhosis of kidney, 157
    in scarlatinal nephritis, 204
Albuminuric retinitis, 164
Alcohol as a cause of chronic nephritis, 112
    of albuminuria, 52
Alcoholism distinguished from uræmic coma, 73
Alkaline treatment of acute nephritis, 104
    of scarlatinal nephritis, 219
Alkaloids in urine as sources of error in albumin testing, 37
Alterations in vascular walls as a cause of albuminuria, 26
Ammonium carbonate as a cause of uræmia, 59
    hydrochlorate in treatment of retinitis, 188
Amyl nitrite in treatment of puerperal convulsions, 246
Amyloid disease of kidneys, 250
Anæmia in acute nephritis, 95
Antiseptic surgery in lardaceous disease, 268

Anuria in scarlatinal nephritis, 219
  in puerperal nephritis, 238
Apocynum cannabinum in treatment of chronic nephritis, 135
  of cirrhotic kidney, 186
  of puerperal nephritis, 244
Apoplexy, how distinguished from uræmia, 73
Arsenic in treatment of chronic nephritis, 132
Arterial tension, evidences of increase of, 154
  treatment of, in renal cirrhosis, 181
  in scarlatinal nephritis, 208
Artificial delivery in puerperal eclampsia, 247

BATHS in albuminuria, 53
Benzoate of lime in treatment of chronic nephritis, 132
Beverages in albuminuria, 52
Blancard's pills in treatment of lardaceous disease, 270
Bleeding in puerperal eclampsia, 245
Blood, changes of, in acute nephritis, 96
  as a cause of albuminuria, 29
  of puerperal nephritis, 228
  of lardaceous disease, 250
  casts, 34
  pressure as a cause of albuminuria, 24
Bowels, regulation of, in treatment of cirrhosis, 189
  derangement of, in scarlatinal nephritis, 206
Bright's disease, acute, 86
  chronic, 140
Brine test for albumin, 43
Bronchial disorders in scarlatinal nephritis, 209
Broom in treatment of chronic nephritis, 135
Burns as a cause of acute nephritis, 88

CAFFEIN in treatment of chronic nephritis, 133
Calcium chloride in treatment of lardaceous disease, 267
Calomel in treatment of chronic nephritis, 132
Cantharides a cause of acute nephritis, 88
Cardiac hypertrophy in cirrhosis of kidney, 156–160
  dilatation in cirrhosis of kidney, 161
  complications in scarlatinal nephritis, 206
  in puerperal nephritis, 234
Casts, renal, 32
Catarrh of bladder distinguished from albuminuria, 20
Catarrhal nephritis, acute, 86
  chronic, 109
Cheese, dangers of use of, in albuminuria, 50

Chloral hydrate in uræmic convulsions, 82
Chlorate of potassium, dangers of, in scarlatina, 216
Chloride of gold in treatment of chronic nephritis, 137
    of renal cirrhosis, 179
Chronic Bright's disease, 140
    suppuration a cause of lardaceous disease, 251
    uræmia, 62
    nephritis, 109
Circulatory system, derangement of, in acute nephritis, 95
    in cirrhosis of kidney, 154
    in scarlatinal nephritis, 206
    in puerperal nephritis, 234
    in cyanotic kidney, 274
    in lardaceous disease, 260
    in chronic nephritis, 120
Cirrhosis of the kidney, 140
    etiology of, 140
    morbid anatomy of, 146
    symptoms of, 151
    course and duration of, 170
    diagnosis of, 173
    prognosis of, 175
    treatment of, 177
    of liver as a symptom of renal cirrhosis, 169
Citrate of potassium in treatment of renal cirrhosis, 178
    of puerperal nephritis, 244
Climate in treatment of albuminuria, 56
    as a cause of renal cirrhosis, 145
Cobalt yellow as a tension depressant, 182
Cold as a cause of acute nephritis, 86
    of chronic nephritis, 110
    of scarlatinal nephritis, 195
    baths, dangers of, in albuminuria, 54
Collapse in scarlatinal nephritis, 207
Coma, uræmic, 71
Concentrated saline cathartics in treatment of cirrhosis, 187
Constipation, dangers arising from, in scarlatina, 216
Convallaria majallis in treatment of renal cirrhosis, 186
    of scarlatinal nephritis, 223
    of cyanotic kidney, 279
Convulsions, uræmic, 70
    puerperal, 236
Copaiba in treatment of chronic nephritis, 132
Copper sulphate in diarrhœa of lardaceous disease, 269
Corrosive sublimate in treatment of chronic nephritis, 132

Counter-irritation in scarlatinal nephritis, 222
Croton oil in treatment of scarlatinal nephritis, 222
Croupous nephritis, 86
Cutaneous system in renal cirrhosis, 163
Cyanotic induration of kidney, 271
    etiology of, 271
    morbid anatomy of, 272
    symptoms of, 273
    course and duration of, 275
    diagnosis of, 277
    prognosis of, 278
    treatment of, 278

DEAFNESS in uræmia, 68
Delirium of uræmia, 69
Diarrhœa in renal cirrhosis, 188
    in uræmia, 67
    in lardaceous disease, 260
    in scarlatinal nephritis, 221
Diet in albuminuria, 49
    in acute nephritis, 102
Diffuse nephritis, chronic, 109
Digestive system, disorders of, in uræmia, 67
    in acute nephritis, 95
    in chronic nephritis, 121
    in renal cirrhosis, 152, 158
    in scarlatinal nephritis, 205
    in puerperal nephritis, 234
    in lardaceous disease, 259
Digitalis in treatment of acute nephritis, 103
    of early renal cirrhosis, 178
    of advanced renal cirrhosis, 184
    of scarlatinal nephritis, 217, 222
    of puerperal nephritis, 244
    of cyanotic kidney, 279
Dilatation of heart in renal cirrhosis, 161
    in scarlatinal nephritis, 207
Disorders of vision in renal cirrhosis, 164
Distilled water as a beverage in albuminuria, 53
Diuretics in treatment of uræmia, 76
    of acute nephritis, 103
    of chronic nephritis, 134
    of puerperal nephritis, 244
Dropsy in acute nephritis, 94
    in chronic nephritis, 118

Dropsy in cirrhosis of kidney, 169
    in scarlatinal nephritis, 204
    in puerperal nephritis, 232
    in lardaceous disease, 258
    in cyanotic induration of kidney, 274
Dyspepsia as a cause of renal cirrhosis, 143
Dyspnœa in renal cirrhosis, 159
    in cyanotic kidney, 274

EARLY stage of renal cirrhosis, treatment of, 177
Eclampsia, puerperal, 236
Eggs, dangers of, as a diet in albuminuria, 50
Elaterium, in treatment of uræmia, 80
Endocarditis in scarlatinal nephritis, 208
Epilepsy distinguished from uræmic convulsions, 72
Epistaxis in early cirrhosis of kidney, 156
Epithelial casts, 33
Epithelium, degeneration of, as a cause of albuminuria, 28
Ergot in treatment of renal cirrhosis, 187
    of scarlatinal nephritis, 218
Excretory products of fœtus as a cause of puerperal nephritis, 229
Exercise in albuminuria, 55
    dangers of, in certain stages of renal cirrhosis, 180

FALSE albuminuria, 18
Farinaceous diet in albuminuria, 52
Ferrocyanide of potassium test for albumin, 42
Fever as a cause of nephritis, 87
Fish as a diet in albuminuria, 51
Fœtal products a cause of puerperal nephritis, 229
Fomentations in treatment of acute nephritis, 105

GALLIC acid in treatment of chronic nephritis, 132
Globulinuria, 22.
Gold chloride in treatment of renal cirrhosis, 179
    of chronic nephritis, 137
Gout as a cause of renal cirrhosis, 143
Gouty kidney, 140
Granular degeneration of kidney, 140

HABITS as a cause of renal cirrhosis, 141
Headache in uræmia, 67
    in puerperal nephritis, 237

Heart changes in renal cirrhosis, 156
  acute hypertrophy of, in scarlatinal nephritis, 206
    dilatation of, in scarlatinal nephritis, 207
    changes of, in puerperal nephritis, 234
Heat a cause of albuminuria, 25
  test for albumin in urine, 37
Hemorrhages, treatment of, in renal cirrhosis, 184
Heredity as a cause of renal cirrhosis, 144
High arterial tension, sphygmographic indications of, 155
  in scarlatinal nephritis, 208
Hot baths in uræmia, 78
  in acute nephritis, 105
  in renal cirrhosis, 186
  contraindications of, in renal cirrhosis, 183
Hypertrophy of heart in renal cirrhosis, 156, 160
  in puerperal nephritis, 235
  in scarlatinal nephritis, 206
Hypodermatic use of morphine in puerperal convulsions, 245

INDUCTION of premature labor in eclampsia, 247
Inflammation as a cause of albuminuria, 30
Insomnia in uræmia, 68
Interstitial nephritis, 140
Intoxication, uræmic, 69
Iodoform in treatment of lardaceous disease, 267
Iron, subcarbonate of, in uræmia, 77
  in treatment of acute nephritis, 108
  of chronic nephritis, 129
  of late stage of renal cirrhosis, 185
  of lardaceous disease of kidneys, 270
Itching of skin in uræmia, 67

JABORANDI in treatment of uræmia, 79
  of renal cirrhosis, 188
Juniper in treatment of uræmia, 76

KIDNEYS, acute inflammation of, 86
  chronic inflammation of, 109
  cirrhosis of, 140
  lardaceous degeneration of, 250
  cyanotic induration of, 271

LARDACEOUS degeneration of kidneys, 250
  etiology of, 250

Lardaceous degeneration of kidneys, morbid anatomy of, 252
    symptoms of, 256
    diagnosis of, 264
    prognosis of, 265
    treatment of, 266
Lead as a cause of renal cirrhosis, 145
    acetate in treatment of diarrhœa of lardaceous disease, 269
Leeching in treatment of acute nephritis, 105
    of scarlatinal nephritis, 221

MALARIA as a cause of chronic nephritis, 111
    of cirrhosis of kidneys, 145
Meat, dangers of, as a diet in albuminuria, 51
Meningitis, relations of, to uræmic convulsions, 60
Menstruation, influence of, upon albuminuria, 55
Mercury, protoiodide of, in treatment of renal cirrhosis, 179
    bichloride of, in treatment of renal cirrhosis, 179
Microörganisms as a cause of scarlatinal nephritis, 195
Milk diet in albuminuria, 51
    in acute nephritis, 108
    as a prophylactic against scarlatinal nephritis, 216
    in puerperal nephritis, 243
    in chronic nephritis, 130
Morphine, hypodermatic use of, in puerperal convulsions, 245
    in uræmic convulsions, 83
Mucin a source of error in testing for albumin in urine, 36
Muscarine as an eliminant in uræmia, 81
Muscular system, derangements of, in chronic nephritis, 121
    in renal cirrhosis, 162

NEPHRITIS, acute, 86
    etiology of, 86
    morbid anatomy of, 89
    symptoms of, 93
    course and duration of, 97
    diagnosis of, 99
    prognosis of, 101
    treatment of, 102
  chronic, 109
    etiology of, 109
    morbid anatomy of, 112
    symptoms of, 117
    course and duration of, 123
    diagnosis of, 124
    prognosis of, 125

Nephritis, chronic, treatment of, 127
  puerperal, 226
  scarlatinal, 192
Nervous system, disorders of, in acute nephritis, 96
  in chronic nephritis, 121
  in cirrhosis of kidneys, 163
  in scarlatinal nephritis, 209
  in puerperal nephritis, 235
  in uræmia, 68
Nitric acid test for albumin, 39
Nitrites in treatment of acute nephritis, 106
  in treatment of renal cirrhosis, 181
Nitrite of amyl in treatment of puerperal eclampsia, 246
  of sodium in treatment of renal cirrhosis, 181
  of potassium in treatment of renal cirrhosis, 181
Nitroglycerine in treatment of renal cirrhosis, 181
  of puerperal convulsions, 246
Non-albuminuric stages of renal cirrhosis, 157

Œdema of the brain as a cause of uræmia, 60
Oleo-resins a source of error in testing for albumin, 37
Opium, dangers incurred by its use in renal cirrhosis, 188
Oxygen in treatment of puerperal convulsions, 247

Paper tests for albumin in urine, 45
Paralysis of heart in scarlatinal nephritis, 207
  in scarlatinal nephritis, 210
Parenchymatous nephritis, acute, 86
  chronic, 109
Peptonuria, 23
Peptones in urine a source of error in albumin testing, 37
Pericarditis in scarlatinal nephritis, 208
Phosphates in urine a source of error in albumin testing, 36
Phosphorus a cause of nephritis, 88
Picric acid test for albumin, 40
Pilocarpin in treatment of acute nephritis, 106
Plumbism as a cause of renal cirrhosis, 145
Pneumonia in scarlatinal nephritis, 209
Potassio-cobaltic nitrite as a tension depressant, 182
Potassio-mercuric iodide test for albumin, 41
Potassium salts in treatment of scarlatinal nephritis, 220
  of chronic nephritis, 135
  of acute nephritis, 104
  of renal cirrhosis, 181
  in the blood a cause of uræmia, 60

INDEX.   293

Potassium iodide in treatment of renal cirrhosis, 179
    of lardaceous disease, 267
Pregnancy as a cause of chronic nephritis, 110
Pressure as a cause of puerperal nephritis, 227
Prophylactic treatment of scarlatinal nephritis, 215
Prostatic discharges distinguished from albumin, 21
Protoiodide of mercury in treatment of renal cirrhosis, 179
Psychic influences over albuminuria, 56
    as a cause of renal cirrhosis, 142
Puerperal nephritis, 226
    etiology of, 227
    morbid anatomy of, 230
    symptoms of, 231
    diagnosis of, 240
    prognosis of, 241
    treatment of, 243
Pulmonary œdema in scarlatinal nephritis, 208
Pulse, characters of, in cirrhosis of kidneys, 154
Pupils in uræmic coma, 71
Purgatives in treatment of uræmia, 80
Pus, characters of, in urine, 19
Pyelitis distinguished from albuminuria, 21
Pyrexia as a cause of albuminuria, 30
    in scarlatinal nephritis, 202

QUANTITATIVE testing for albumin, 47
Quantity of urine in uræmia, 63
Quinine in treatment of scarlatinal nephritis, 222

RENAL casts, 32
    cirrhosis, 140
    veins, pressure upon, as a cause of puerperal nephritis, 227
Resorts for albuminuric patients, 57
Respiratory system, disorders of, in acute nephritis, 97
        in renal cirrhosis, 159
        in scarlatinal nephritis, 208
        in cyanotic kidney, 274
        in lardaceous disease, 262
Rules for estimating solids of the urine, 64

SALICYLATES, dangers from use of, in scarlatina, 216
Saline purgatives in treatment of uræmia, 81
    of renal cirrhosis, 187
Sandal-wood oil in treatment of chronic nephritis, 132
Scarlatina as a cause of chronic nephritis, 109

Scarlatinal nephritis, 192
    etiology of, 194
    morbid anatomy of, 196
    symptoms of, 202
    course and duration of, 210
    diagnosis of, 212
    prognosis of, 214
    treatment of, 215
Significance of albuminuria, 31
Skin, the care of, in albuminuria, 53
    lesions of, as a cause of acute nephritis, 88
Sodium nitrite in treatment of renal cirrhosis, 181
    salts in treatment of scarlatinal nephritis, 220
    tannate in treatment of chronic nephritis, 132
    tungstate test for albumin, 43
    sulphate, effects of, when injected into the veins, 59
Solids of the urine in uræmia, 64
Sphygmograms, characters of, in renal cirrhosis, 155
Squill, as a diuretic, in treatment of uræmia, 76
    in treatment of chronic nephritis, 132
Stage of cardiac enlargement in renal cirrhosis, treatment of, 180
Stomach disorders in uræmia, 67
Strophanthus in treatment of cirrhosis of kidneys, 187
    of scarlatinal nephritis, 223
    of cyanotic kidneys, 279
Strychnia in treatment of renal cirrhosis, 184
Sulphate of copper in treatment of diarrhœa of lardaceous disease, 269
Suppuration as a cause of chronic nephritis, 111
    of lardaceous disease, 251
Syphilis as a cause of chronic nephritis, 111
    of lardaceous disease, 252

TANNATE of sodium in treatment of chronic nephritis, 132
Tar in treatment of chronic nephritis, 132
Toxic causes of acute nephritis, 88
Tubal nephritis, chronic, 109
Tuberculosis as a cause of lardaceous disease, 251
Turpentine as a cause of acute nephritis, 88

URÆMIA, 58
    causes of, 58
    symptoms of, 62
    diagnosis of, 72
    prognosis of, 74
    treatment of, 75

Uræmia in scarlatinal nephritis, 209
Uræmic amaurosis, 164
Urates in urine a source of error in testing for albumin, 36
Urea as a cause of uræmia, 58
    quantitative estimation of, 65
Urine, characters of, in acute nephritis, 93
    in chronic nephritis, 117
    in cirrhosis of kidneys, 153, 158
    in scarlatinal nephritis, 202
    in puerperal nephritis, 232
    in cyanotic kidney, 275
    in lardaceous disease of kidneys, 257

VASCULAR system in renal cirrhosis, 160
Venesection in uræmic convulsions, 82
    in scarlatinal nephritis, 221
    in puerperal eclampsia, 245
Vertigo in renal cirrhosis, 152
Vision, disturbances of, in uræmia, 68
Visual disorders in scarlatinal nephritis, 209
    in renal cirrhosis, 164
    in puerperal nephritis, 237
Vomiting in uræmia, 67
    in scarlatina, 205
    in puerperal nephritis, 234

WOOL garments, importance of use of, in albuminuria, 54

LEA BROTHERS & CO.'S

(Late HENRY C. LEA'S SON & CO.)

CLASSIFIED CATALOGUE

OF

MEDICAL AND SURGICAL
PUBLICATIONS.

In asking the attention of the profession to the works advertised in the following pages, the publishers would state that no pains are spared to secure a continuance of the confidence earned for the publications of the house by their careful selection and accuracy and finish of execution.

The large number of inquiries received from the profession for a finer class of bindings than is usually placed on medical books has induced us to put certain of our standard publications in half Russia; and, that the growing taste may be encouraged, the prices have been fixed at so small an advance over the cost of sheep as to place it within the means of all to possess a library that shall have attractions as well for the eye as for the mind of the reading practitioner.

The printed prices are those at which books can generally be supplied by booksellers throughout the United States, who can readily procure for their customers any works not kept in stock. Where access to bookstores is not convenient books will be sent by mail postpaid on receipt of the price, and as the limit of mailable weight has been removed, no difficulty will be experienced in obtaining through the post-office any work in this catalogue. No risks, however, are assumed either on the money or on the books, and no publications but our own are supplied, so that gentlemen will in most cases find it more convenient to deal with the nearest bookseller.

LEA BROTHERS & CO.

Nos. 706 and 708 Sansom St., Philadelphia. May, 1886.

PROSPECTUS FOR 1886.
The American Journal of the Medical Sciences.

Quarterly, 300-350 pages, with illustrations. Price, $5.00 per annum.

With the year 1886 THE AMERICAN JOURNAL OF THE MEDICAL SCIENCES became in Great Britain the recognized organ of the profession—a position similar to that occupied by it in America for sixty-six years. On its announcement, this project for an international journal was welcomed abroad with acclamation, and one hundred and thirty-five of the foremost English practitioners authorized the use of their names as contributors in order to aid in extending over their country the benefits which American medicine has enjoyed from the existence of THE JOURNAL during two generations. This friendly challenge was accepted with an almost equal number of Americans, to whose proved ability this country can well afford to entrust her reputation.

In thus becoming the medium of communication between the two nations distinguished above all others by the practical character of their labors, THE JOURNAL will undoubtedly form the most efficient factor in medical progress which the world has yet seen. Already this generous spirit of rivalry has proved that the ample space devoted to *Original Articles* will be filled with a series of contributions unapproachable in value.

THE AMERICAN JOURNAL of the MEDICAL SCIENCES.
(Continued from first page.)

But it is not only in the *Original Department* that the JOURNAL of the future will seek to eclipse all its efforts in the past. The mass of contributions to medical literature and science increases with such rapidity, that if the reader is to keep abreast with them the matter must be carefully sifted, and arranged so as to enable him to grasp it understandingly with the least possible expenditure of time. In the *Bibliographical Department* therefore, separate reviews will be devoted only to works of exceptional importance. As a rule, new books will be considered in groups of cognate subjects, the reviewer setting forth tersely the merits of the individual volumes with a condensed statement of the views of the authors. In this manner the reader will be kept advised of the products of the press in the most convenient manner.

A similar plan will be adopted in the *Quarterly Summary of Progress*. The various branches of medical science have been assigned to the following gentlemen, who will furnish well-digested *résumés* of progress, paying special attention to clinical application : Anatomy, George D. Thane, M. R. C. S.; Physiology, Gerald F. Yeo, M. D.; Materia Medica, Therapeutics and Pharmacology, Roberts Bartholow, M. D., LL. D.; Surgery, in America, Lewis A. Stimson, M. D.; in Europe, Frederick Treves, F. R. C. S.; Ophthalmology, Charles S. Bull, M. D.; Otology, Charles H. Burnett, M. D.; Laryngology, J. Solis Cohen, M. D.; Dermatology, Louis A. Duhring, M. D., and H. W. Stelwagon, M. D.; Midwifery and Gynæcology, Francis H. Champneys, M. B., F. R. C. P.; Jurisprudence. Matthew Hay, M. D.; Public Health, Shirley F. Murphy, M. R. C. S.

The publishers feel an honest pride in thus being the means of bringing together the professions of the two great English speaking peoples, and in laying before them a periodical which must be universally recognized as marking a new era in medical progress. Believing that it will be regarded as indispensable by all intelligent physicians on both sides of the Atlantic, they feel themselves warranted, by the expectation of a large increase in circulation, in maintaining the present very moderate subscription price, notwithstanding the greatly augmented expenditure entailed by the change.

The JOURNAL will continue to be published quarterly, as heretofore, on the first of January, April, July and October.

Price, FIVE DOLLARS Per Annum, in Advance.

THE MEDICAL NEWS.
A National Weekly Periodical, containing 28 to 32 Quarto Pages in Each Issue.

The continually increasing appreciation of THE MEDICAL NEWS by the profession throughout the country, encourages its managers to make during the coming year even more strenuous efforts to enhance its value to the practitioner.

Possessing a most efficient organization THE NEWS unites the best features of the medical magazine and newspaper. Its large and able Editorial Staff discusses in each issue the important topics of the day in a thoughtful and scholarly manner, while its corps of qualified reporters and correspondents, covering every medical centre, insures that its readers shall be promptly and thoroughly posted upon all matters of interest in the world of medicine. On account of the position conceded to The News, it has become the medium chosen by the leading minds of the profession for the publication of their most important contributions to medical science. The valuable instruction afforded in Clinical Lectures, and the rich experience gained in the leading Hospitals of the world are constantly laid before the readers of The News, while prompt and authentic reports of Society Proceedings are received from special reporters in various sections of the country by mail and Telegraph. In the pages devoted to the progress of Medical Science are found early notes of all important advances, gleaned from the principal journals of both hemispheres. Ample space is devoted to Reviews, News Items, Correspondence and Notes and Queries In short, every branch of medicine is adequately represented in The News, and the details of plan and typography have been carefully studied in order to economize the time and secure the comfort of the reader in every possible way.

Price, FIVE DOLLARS Per Annum, in Advance.

COMMUTATION RATE.
To subscribers paying in advance for 1886 :—
AMERICAN JOURNAL OF THE MEDICAL SCIENCES (quarterly) } To one address for $9.00
THE MEDICAL NEWS (weekly) } per annum.

LEA BROTHERS & CO.'S PUBLICATIONS—Period., Manuals. 3

SPECIAL OFFER.

THE YEAR-BOOK OF TREATMENT for 1885 (for full description of which see page 17), will be sent to any subscriber to either of the above periodicals, free by mail, on receipt of 75 cents (regular price $1.25).

Subscribers can obtain, at the close of each volume, cloth covers for THE JOURNAL *(one annually), and for* THE NEWS *(one annually), free by mail, by remitting Ten Cents for the* JOURNAL *cover, and Fifteen Cents for the* NEWS *cover.*

☞ The safest mode of remittance is by bank check or postal money order, drawn to the order of the undersigned; where these are not accessible, remittances for subscriptions may be sent at the risk of the publishers by forwarding in *registered* letters. Address,

LEA BROTHERS & CO., 706 and 708 Sansom Street, PHILADELPHIA.

THE MEDICAL NEWS VISITING LIST FOR 1887.

Containing Calendar for two years. Obstetric diagrams. Scheme of Dentition. Tables of weights and measures and Comparative scales. Instructions for examining the urine. List of disinfectants. Table of eruptive fevers. Lists of remedies not generally used, and Incompatibles, Poisons and Antidotes. Artificial respiration. Table of doses, prepared to accord with the last revision of the U. S. Pharmacopœia, an extended table of Diseases and their remedies, and directions for Ligation of Arteries. Handsomely bound in limp Morocco, with tuck and pencil and rubber. *Preparing.*

Issued towards the close of 1885 as an experiment, the large edition of *The Medical News Visiting List* for 1886 was rapidly exhausted. The great favor thus shown has placed this venture upon a permanent basis, and the profession may therefore look in due time for *The List* for 1887, in full confidence that it will contain such new features as may, from time to time, prove themselves desirable and possible. In response to numerous requests, *The List* will in future be issued in several different sizes to accommodate all the records of practices varying from thirty to ninety patients per week.

THE MEDICAL NEWS PHYSICIANS' LEDGER.

Containing 400 pages of fine linen "ledger" paper, ruled so that all the accounts of a large practice may be conveniently kept in it, either by single or double entry, for a long period. Strongly bound in leather, with cloth sides, and with a patent flexible back, which permits it to lie perfectly flat when opened at any place. Price, $5.00. Also a small special lot of same Ledger, with 300 pages. Price, $4.00.

HARTSHORNE, HENRY, A. M., M. D., LL. D.,
Lately Professor of Hygiene in the University of Pennsylvania.

A Conspectus of the Medical Sciences; Containing Handbooks on Anatomy, Physiology, Chemistry, Materia Medica, Practice of Medicine, Surgery and Obstetrics. Second edition, thoroughly revised and greatly improved. In one large royal 12mo. volume of 1028 pages, with 477 illustrations. Cloth, $4.25; leather, $5.00.

The object of this manual is to afford a convenient work of reference to students during the brief moments at their command while in attendance upon medical lectures. It is a favorable sign that it has been found necessary, in a short space of time, to issue a new and carefully revised edition. The illustrations are very numerous and unusually clear, and each part seems to have received its due share of attention. We can conceive such a work to be useful, not only to students, but to practitioners as well. It reflects credit upon the industry and energy of its able editor.—*Boston Medical and Surgical Journal*, Sept. 3, 1874.

We can say with the strictest truth that it is the best work of the kind with which we are acquainted. It embodies in a condensed form all recent contributions to practical medicine, and is therefore useful to every busy practitioner throughout our country, besides being admirably adapted to the use of students of medicine. The book is faithfully and ably executed.—*Charleston Medical Journal*, April, 1875.

NEILL, JOHN, M. D., and SMITH, F. G., M. D.,
Late Surgeon to the Penna. Hospital. *Prof. of the Institutes of Med. in the Univ. of Penna.*

An Analytical Compendium of the Various Branches of Medical Science, for the use and examination of Students. A new edition, revised and improved. In one large royal 12mo. volume of 974 pages, with 374 woodcuts. Cloth, $4; leather, $4.75.

LUDLOW, J. L., M. D.,
Consulting Physician to the Philadelphia Hospital, etc.

A Manual of Examinations upon Anatomy, Physiology, Surgery, Practice of Medicine, Obstetrics, Materia Medica, Chemistry, Pharmacy and Therapeutics. To which is added a Medical Formulary. 3d edition, thoroughly revised, and greatly enlarged. In one 12mo. volume of 816 pages, with 370 illustrations. Cloth, $3.25; leather, $3.75.

The arrangement of this volume in the form of question and answer renders it especially suitable for the office examination of students, and for those preparing for graduation.

DUNGLISON, ROBLEY, M. D.,
Late Professor of Institutes of Medicine in the Jefferson Medical College of Philadelphia.

MEDICAL LEXICON; A Dictionary of Medical Science: Containing a concise Explanation of the various Subjects and Terms of Anatomy, Physiology, Pathology, Hygiene, Therapeutics, Pharmacology, Pharmacy, Surgery, Obstetrics, Medical Jurisprudence and Dentistry, Notices of Climate and of Mineral Waters, Formulæ for Officinal, Empirical and Dietetic Preparations, with the Accentuation and Etymology of the Terms, and the French and other Synonymes, so as to constitute a French as well as an English Medical Lexicon. Edited by RICHARD J. DUNGLISON, M. D. In one very large and handsome royal octavo volume of 1139 pages. Cloth, $6.50; leather, raised bands, $7.50; very handsome half Russia, raised bands, $8.

The object of the author, from the outset, has not been to make the work a mere lexicon or dictionary of terms, but to afford under each word a condensed view of its various medical relations, and thus to render the work an epitome of the existing condition of medical science. Starting with this view, the immense demand which has existed for the work has enabled him, in repeated revisions, to augment its completeness and usefulness, until at length it has attained the position of a recognized and standard authority wherever the language is spoken. Special pains have been taken in the preparation of the present edition to maintain this enviable reputation. The additions to the vocabulary are more numerous than in any previous revision, and particular attention has been bestowed on the accentuation, which will be found marked on every word. The typographical arrangement has been greatly improved, rendering reference much more easy, and every care has been taken with the mechanical execution. The volume now contains the matter of at least four ordinary octavos.

About the first book purchased by the medical student is the Medical Dictionary. The lexicon explanatory of technical terms is simply a *sine qua non*. In a science so extensive and with such collaterals as medicine, it is as much a necessity also to the practising physician. To meet the wants of students and most physicians the dictionary must be condensed while comprehensive, and practical while perspicacious. It was because Dunglison's met these indications that it became at once the dictionary of general use wherever medicine was studied in the English language. In no former revision have the alterations and additions been so great. The chief terms have been set in black letter, while the derivatives follow in small caps; an arrangement which greatly facilitates reference. —*Cincinnati Lancet and Clinic*, Jan. 10, 1874.

A book of which every American ought to be proud. When the learned author of the work passed away, probably all of us feared lest the book should not maintain its place in the advancing science whose terms it defines. Fortunately, Dr. Richard J. Dunglison, having assisted his father in the revision of several editions of the work, and having been, therefore, trained in the methods and imbued with the spirit of the book, has been able to edit it as a work of the kind should be edited—to carry it on steadily, without jar or interruption, along the grooves of thought it has travelled during its lifetime. To show the magnitude of the task which Dr. Dunglison has assumed and carried through, it is only necessary to state that more than six thousand new subjects have been added in the present edition.—*Philadelphia Medical Times*, Jan. 3, 1874.

It has the rare merit that it certainly has no rival in the English language for accuracy and extent of references.—*London Medical Gazette*.

HOBLYN, RICHARD D., M. D.
A Dictionary of the Terms Used in Medicine and the Collateral Sciences. Revised, with numerous additions, by ISAAC HAYS, M. D., late editor of The American Journal of the Medical Sciences. In one large royal 12mo. volume of 520 double-columned pages. Cloth, $1.50; leather, $2.00.

It is the best book of definitions we have, and ought always to be upon the student's table.—*Southern Medical and Surgical Journal.*

STUDENTS' SERIES OF MANUALS.
A Series of Fifteen Manuals, for the use of Students and Practitioners of Medicine and Surgery, written by eminent Teachers or Examiners, and issued in pocket-size 12mo. volumes of 300-540 pages, richly illustrated and at a low price. The following volumes are now ready: BELL'S *Comparative Physiology and Anatomy*, GOULD'S *Surgical Diagnosis*, ROBERTSON'S *Physiological Physics*, BRUCE'S *Materia Medica and Therapeutics*, POWER'S *Human Physiology*, CLARKE and LOCKWOOD'S *Dissectors' Manual*, RALFE'S *Clinical Chemistry*, TREVES' *Surgical Applied Anatomy*, PEPPER'S *Surgical Pathology*, and KLEIN'S *Elements of Histology*. The following are in press: BELLAMY'S *Operative Surgery*, PEPPER'S *Forensic Medicine*, and CURNOW'S *Medical Applied Anatomy*. For separate notices see index on last page.

SERIES OF CLINICAL MANUALS.
In arranging for this Series it has been the design of the publishers to provide the profession with a collection of authoritative monographs on important clinical subjects in a cheap and portable form. The volumes will contain about 550 pages and will be freely illustrated by chromo-lithographs and woodcuts. The following volumes are now ready: TREVES' *Manual of Surgery*, by various writers, in three volumes; OWEN on *Surgical Diseases of Children*, MORRIS on *Surgical Diseases of the Kidney*, PICK on *Fractures and Dislocations*, BUTLIN on the *Tongue*, TREVES on *Intestinal Obstruction*, and SAVAGE on *Insanity and Allied Neuroses*. The following are in active preparation: HUTCHINSON on *Syphilis*, BRYANT on the *Breast*, BROADBENT on the *Pulse*, LUCAS on *Diseases of the Urethra*, MARSH on *Diseases of the Joints*, and BALL on the *Rectum and Anus*. For separate notices see index on last page.

GRAY, HENRY, F. R. S.,
Lecturer on Anatomy at St. George's Hospital, London.

Anatomy, Descriptive and Surgical. The Drawings by H. V. CARTER, M. D., and Dr. WESTMACOTT. The dissections jointly by the AUTHOR and Dr. CARTER. With an Introduction on General Anatomy and Development by T. HOLMES, M. A., Surgeon to St. George's Hospital. Edited by T. Pickering Pick, F. R. C. S., Surgeon to and Lecturer on Anatomy at St. George's Hospital, London, Examiner in Anatomy, Royal College of Surgeons of England. A new American from the tenth enlarged and improved London edition. To which is added the second American from the latest English edition of LANDMARKS, MEDICAL AND SURGICAL, by LUTHER HOLDEN, F. R. C. S., author of "Human Osteology," "A Manual of Dissections," etc. In one imperial octavo volume of 1023 pages, with 564 large and elaborate engravings on wood. Cloth, $6.00; leather, $7.00; very handsome half Russia, raised bands, $7.50.

This work covers a more extended range of subjects than is customary in the ordinary text-books, giving not only the details necessary for the student, but also the application to those details to the practice of medicine and surgery. It thus forms both a guide for the learner and an admirable work of reference for the active practitioner. The engravings form a special feature in the work, many of them being the size of nature, nearly all original, and having the names of the various parts printed on the body of the cut, in place of figures of reference with descriptions at the foot. They thus form a complete and splendid series, which will greatly assist the student in forming a clear idea of the parts, and will also serve to refresh the memory of those who may find in the exigencies of practice the necessity of recalling the details of the dissecting-room. Combining, as it does, a complete Atlas of Anatomy with a thorough treatise on systematic, descriptive and applied Anatomy, the work will be found of great service to all physicians who receive students in their offices, relieving both preceptor and pupil of much labor in laying the groundwork of a thorough medical education.

Landmarks, Medical and Surgical, by the distinguished Anatomist, Mr. Luther Holden, has been appended to the present edition as it was to the previous one. This work gives in a clear, condensed and systematic way all the information by which the practitioner can determine from the external surface of the body the position of internal parts. Thus complete, the work, it is believed, will furnish all the assistance that can be rendered by type and illustration in anatomical study.

This well-known work comes to us as the latest American from the tenth English edition. As its title indicates, it has passed through many hands and has received many additions and revisions. The work is not susceptible of more improvement. Taking it all in all, its size, manner of make-up, its character and illustrations, its general accuracy of description, its practical aim, and its perspicuity of style, it is here the Anatomy best adapted to the wants of the student and practitioner.—*Medical Record*, Sept. 15, 1883.

There is probably no work used so universally by physicians and medical students as this one. It has received the confidence that they repose in it. If the present edition is compared with that issued two years ago, one will readily see how much it has been improved in that time. Many pages have been added to the text, especially in those parts that treat of histology, and many new cuts have been introduced and old ones modified.—*Journal of the American Medical Association*, Sept. 1, 1883.

ALSO FOR SALE SEPARATE—

HOLDEN, LUTHER, F. R. C. S.,
Surgeon to St. Bartholomew's and the Foundling Hospitals, London.

Landmarks, Medical and Surgical. Second American from the latest revised English edition, with additions by W. W. KEEN, M. D., Professor of Artistic Anatomy in the Pennsylvania Academy of the Fine Arts, formerly Lecturer on Anatomy in the Philadelphia School of Anatomy. In one handsome 12mo. volume of 148 pages. Cloth, $1.00.

This little book is all that can be desired within its scope, and its contents will be found simply invaluable to the young surgeon or physician, since they bring before him such data as he requires at every examination of a patient. It is written in language so clear and concise that one ought almost to learn it by heart. It teaches diagnosis by external examination, ocular and palpable, of the body, with such anatomical and physiological facts as directly bear on the subject. It is eminently the student's and young practitioner's book.—*Physician and Surgeon*, Nov. 1881.

The study of these Landmarks by both physicians and surgeons is much to be encouraged. It inevitably leads to a progressive education of both the eye and the touch, by which the recognition of disease or the localization of injuries is vastly assisted. One thoroughly familiar with the facts here taught is capable of a degree of accuracy and a confidence of certainty which is otherwise unattainable. We cordially recommend the Landmarks to the attention of every physician who has not yet provided himself with a copy of this useful, practical guide to the correct placing of all the anatomical parts and organs.—*Canada Medical and Surgical Journal*, Dec. 1881.

WILSON, ERASMUS, F. R. S.

A System of Human Anatomy, General and Special. Edited by W. H. GOBRECHT, M. D., Professor of General and Surgical Anatomy in the Medical College of Ohio. In one large and handsome octavo volume of 616 pages, with 397 illustrations. Cloth, $4.00; leather, $5.00.

CLELAND, JOHN, M. D., F. R. S.,
Professor of Anatomy and Physiology in Queen's College, Galway.

A Directory for the Dissection of the Human Body. In one 12mo. volume of 178 pages. Cloth, $1.25.

6 Lea Brothers & Co.'s Publications—Anatomy.

ALLEN, HARRISON, M. D.,
Professor of Physiology in the University of Pennsylvania.

A System of Human Anatomy, Including Its Medical and Surgical Relations. For the use of Practitioners and Students of Medicine. With an Introductory Section on Histology. By E. O. SHAKESPEARE, M. D., Ophthalmologist to the Philadelphia Hospital. Comprising 818 double-columned quarto pages, with 380 illustrations on 109 full page lithographic plates, many of which are in colors, and 241 engravings in the text. In six Sections, each in a portfolio. Section I. HISTOLOGY. Section II. BONES AND JOINTS. Section III. MUSCLES AND FASCIÆ. Section IV. ARTERIES, VEINS AND LYMPHATICS. Section V. NERVOUS SYSTEM. Section VI. ORGANS OF SENSE, OF DIGESTION AND GENITO-URINARY ORGANS, EMBRYOLOGY, DEVELOPMENT, TERATOLOGY, SUPERFICIAL ANATOMY, POST-MORTEM EXAMINATIONS, AND GENERAL AND CLINICAL INDEXES. Price per Section, $3.50; also bound in one volume, cloth, $23.00; very handsome half Russia, raised bands and open back, $25.00. *For sale by subscription only. Apply to the Publishers.*

Extract from Introduction.

It is the design of this book to present the facts of human anatomy in the manner best suited to the requirements of the student and the practitioner of medicine. The author believes that such a book is needed, inasmuch as no treatise, as far as he knows, contains, in addition to the text descriptive of the subject, a systematic presentation of such anatomical facts as can be applied to practice.

A book which will be at once accurate in statement and concise in terms; which will be an acceptable expression of the present state of the science of anatomy; which will exclude nothing that can be made applicable to the medical art, and which will thus embrace all of surgical importance, while omitting nothing of value to clinical medicine,—would appear to have an excuse for existence in a country where most surgeons are general practitioners, and where there are few general practitioners who have no interest in surgery.

It is to be considered a study of applied anatomy in its widest sense—a systematic presentation of such anatomical facts as can be applied to the practice of medicine as well as of surgery. Our author is concise, accurate and practical in his statements, and succeeds admirably in infusing an interest into the study of what is generally considered a dry subject. The department of Histology is treated in a masterly manner, and the ground is travelled over by one thoroughly familiar with it. The illustrations are made with great care, and are simply superb. There is as much of practical application of anatomical points to the every-day wants of the medical clinician as to those of the operating surgeon. In fact, few general practitioners will read the work without a feeling of surprised gratification that so many points, concerning which they may never have thought before are so well presented for their consideration. It is a work which is destined to be the best of its kind in any language.—*Medical Record,* Nov. 25, 1882.

CLARKE, W. B., F.R.C.S. & LOCKWOOD, C. B., F.R.C.S.
Demonstrators of Anatomy at St. Bartholomew's Hospital Medical School, London.

The Dissector's Manual. In one pocket-size 12mo. volume of 396 pages, with 49 illustrations. Limp cloth, red edges, $1.50. See *Students' Series of Manuals,* page 4.

This is a very excellent manual for the use of the student who desires to learn anatomy. The methods of demonstration seem to us very satisfactory. There are many woodcuts which, for the most part, are good and instructive. The book is neat and convenient. We are glad to recommend it.—*Boston Medical and Surgical Journal,* Jan. 17, 1884.

TREVES, FREDERICK, F. R. C. S.,
Senior Demonstrator of Anatomy and Assistant Surgeon at the London Hospital.

Surgical Applied Anatomy. In one pocket-size 12mo. volume of 540 pages, with 61 illustrations. Limp cloth, red edges, $2.00. See *Students' Series of Manuals,* page 4.

He has produced a work which will command a larger circle of readers than the class for which it was written. This union of a thorough, practical acquaintance with these fundamental branches, quickened by daily use as a teacher and practitioner, has enabled our author to prepare a work which it would be a most difficult task to excel.—*The American Practitioner* Feb. 1884.

CURNOW, JOHN, M. D., F. R. C. P.,
Professor of Anatomy at King's College, Physician at King's College Hospital.

Medical Applied Anatomy. In one pocket-size 12mo. volume. *Preparing.* See *Students' Series of Manuals,* page 4.

BELLAMY, EDWARD, F. R. C. S.,
Senior Assistant-Surgeon to the Charing-Cross Hospital, London.

The Student's Guide to Surgical Anatomy: Being a Description of the most Important Surgical Regions of the Human Body, and intended as an Introduction to operative Surgery. In one 12mo. volume of 300 pages, with 50 illustrations. Cloth, $2.25.

HARTSHORNE'S HANDBOOK OF ANATOMY AND PHYSIOLOGY. Second edition, revised. In one royal 12mo. volume of 310 pages, with 220 woodcuts. Cloth, $1.75.

HORNER'S SPECIAL ANATOMY AND HISTOLOGY. Eighth edition, extensively revised and modified. In two octavo volumes of 1007 pages, with 320 woodcuts. Cloth, $6.00.

LEA BROTHERS & CO.'S PUBLICATIONS—Physics, Physiol., Anat.   7

DRAPER, JOHN C., M. D., LL. D.,
Professor of Chemistry in the University of the City of New York.
Medical Physics. A Text-book for Students and Practitioners of Medicine. In one octavo volume of 734 pages, with 376 woodcuts, mostly original. Cloth, $4.

From the Preface.

The fact that a knowledge of Physics is indispensable to a thorough understanding of Medicine has not been as fully realized in this country as in Europe, where the admirable works of Desplats and Gariel, of Robertson and of numerous German writers constitute a branch of educational literature to which we can show no parallel. A full appreciation of this the author trusts will be sufficient justification for placing in book form the substance of his lectures on this department of science, delivered during many years at the University of the City of New York.

Broadly speaking, this work aims to impart a knowledge of the relations existing between Physics and Medicine in their latest state of development, and to embody in the pursuit of this object whatever experience the author has gained during a long period of teaching this special branch of applied science.

This elegant and useful work bears ample testimony to the learning and good judgment of the author. He has fitted his work admirably to the exigencies of the situation by presenting the reader with brief, clear and simple statements of such propositions as he is by necessity required to master. The subject matter is well arranged, liberally illustrated and carefully indexed. That it will take rank at once among the text-books is certain, and it is to be hoped that it will find a place upon the shelf of the practical physician, where, as a book of reference, it will be found useful and agreeable.—*Louisville Medical News*, September 26, 1885.

Certainly we have no text-book as full as the excellent one he has prepared. It begins with a statement of the properties of matter and energy. After these the special departments of physics are explained, acoustics, optics, heat, electricity and magnetism, closing with a section on electro-biology. The applications of all these to physiology and medicine are kept constantly in view. The text is amply illustrated and the many difficult points of the subject are brought forward with remarkable clearness and ability.—*Medical and Surgical Reporter*, July 18, 1885. q.

That this work will greatly facilitate the study of medical physics is apparent upon even a mere cursory examination. It is marked by that scientific accuracy which always characterizes Dr. Draper's writings. Its peculiar value lies in the fact that it is written from the standpoint of the medical man. Hence much is omitted that appears in a mere treatise on physical science, while much is inserted of peculiar value to the physician.—*Medical Record*, August 22, 1885.

ROBERTSON, J. McGREGOR, M. A., M. B.,
Muirhead Demonstrator of Physiology, University of Glasgow.
Physiological Physics. In one 12mo. volume of 537 pages, with 219 illustrations. Limp cloth, $2.00. See *Students' Series of Manuals*, page 4.

The title of this work sufficiently explains the nature of its contents. It is designed as a manual for the student of medicine, an auxiliary to his text-book in physiology, and it would be particularly useful as a guide to his laboratory experiments. It will be found of great value to the practitioner. It is a carefully prepared book of reference, concise and accurate, and as such we heartily recommend it.—*Journal of the American Medical Association*, Dec. 6, 1884.

DALTON, JOHN C., M. D.,
Professor Emeritus of Physiology in the College of Physicians and Surgeons, New York.
Doctrines of the Circulation of the Blood. A History of Physiological Opinion and Discovery in regard to the Circulation of the Blood. In one handsome 12mo. volume of 293 pages. Cloth, $2.

Dr. Dalton's work is the fruit of the deep research of a cultured mind, and to the busy practitioner it cannot fail to be a source of instruction. It will inspire him with a feeling of gratitude and admiration for those plodding workers of olden times, who laid the foundation of the magnificent temple of medical science as it now stands.—*New Orleans Medical and Surgical Journal*, Aug. 1885.

In the progress of physiological study no fact was of greater moment, none more completely revolutionized the theories of teachers, than the discovery of the circulation of the blood. This explains the extraordinary interest it has to all medical historians. The volume before us is one of three or four which have been written within a few years by American physicians. It is in several respects the most complete. The volume, though small in size, is one of the most creditable contributions from an American pen to medical history that has appeared.—*Med. & Surg. Rep.*, Dec. 6, 1884.

BELL, F. JEFFREY, M. A.,
Professor of Comparative Anatomy at King's College, London.
Comparative Physiology and Anatomy. In one 12mo. volume of 561 pages, with 229 illustrations. Limp cloth, $2.00. See *Students' Series of Manuals*, page 4.

This is another of the "Students' Series of Manuals," and a most excellent one at that. The descriptions are clear, the illustrations good, and the presswork and paper unexceptionable. The student of biology will be materially benefited by careful investigation of this valuable little work. —*Southern Practitioner*, October, 1885.

ELLIS, GEORGE VINER,
Emeritus Professor of Anatomy in University College, London.
Demonstrations of Anatomy. Being a Guide to the Knowledge of the Human Body by Dissection. From the eighth and revised London edition. In one very handsome octavo volume of 716 pages, with 249 illustrations. Cloth, $4.25; leather, $5.25.

ROBERTS, JOHN B., A. M., M. D.,
Prof. of Applied Anat. and Oper. Surg. in Phila. Polyclinic and Coll. for Graduates in Medicine.
The Compend of Anatomy. For use in the dissecting-room and in preparing for examinations. In one 16mo. volume of 196 pages. Limp cloth, 75 cents.

8   LEA BROTHERS & CO.'S PUBLICATIONS—**Physiology, Chemistry.**

DALTON, JOHN C., M. D.,
Professor of Physiology in the College of Physicians and Surgeons, New York, etc.

A Treatise on Human Physiology. Designed for the use of Students and Practitioners of Medicine. Seventh edition, thoroughly revised and rewritten. In one very handsome octavo volume of 722 pages, with 252 beautiful engravings on wood. Cloth, $5.00; leather, $6.00; very handsome half Russia, raised bands, $6.50.

The merits of Professor Dalton's text-book, his smooth and pleasing style, the remarkable clearness of his descriptions, which leave not a chapter obscure, his cautious judgment and the general correctness of his facts, are perfectly known. They have made his text-book the one most familiar to American students.—*Med. Record,* March 4, 1882.

Certainly no physiological work has ever issued from the press that presented its subject-matter in a clearer and more attractive light. Almost every page bears evidence of the exhaustive revision that has taken place. The material is placed in a

more compact form, yet its delightful charm is re-tained, and no subject is thrown into obscurity. Altogether this edition is far in advance of any previous one, and will tend to keep the profession posted as to the most recent additions to our physiological knowledge.—*Michigan Medical News,* April, 1882.

One can scarcely open a college catalogue that does not have mention of Dalton's *Physiology* as the recommended text or consultation-book. For American students we would unreservedly recommend Dr. Dalton's work.—*Va. Med. Monthly,* July,'82.

FOSTER, MICHAEL, M. D., F. R. S.,
Prelector in Physiology and Fellow of Trinity College, Cambridge, England.

Text-Book of Physiology. Third American from the fourth English edition, with notes and additions by E. T. REICHERT, M. D. In one handsome royal 12mo. volume of 908 pages, with 271 illustrations. Cloth, $3.25; leather, $3.75. *Just ready.*

Dr. Foster's work upon physiology is so well-known as a text-book in this country, that it needs but little be said in regard to it. There is scarcely a medical college in the United States where it is not in the hands of the students. The author, more than any other writer with whom we are acquainted, seems to understand what portions of the science are essential for students

to know and what may be passed over by them as not important. From the beginning to the end, physiology is taught in a systematic manner. To this third American edition numerous additions, corrections and alterations have been made, so that in its present form the usefulness of the book will be found to be much increased.—*Cincinnati Medical News,* July 1885.

POWER, HENRY, M. B., F. R. C. S.,
Examiner in Physiology, Royal College of Surgeons of England.

Human Physiology. In one handsome pocket-size 12mo. volume of 396 pages, with 47 illustrations. Cloth, $1.50. See *Students' Series of Manuals,* page 3.

The prominent character of this work is that of judicious condensation, in which an able and successful effort appears to have been made by its accomplished author to teach the greatest number of facts in the fewest possible words. The result is a specimen of concentrated intellectual pabulum seldom surpassed, which ought to be carefully ingested and digested by every practitioner who desires to keep himself well informed upon this most progressive of the medical sciences. The volume is one which we cordially recommend

to every one of our readers.—*The American Journal of the Medical Sciences,* October, 1884.

This little work is deserving of the highest praise, and we can hardly conceive how the main facts of this science could have been more clearly or concisely stated. The price of the work is such as to place it within the reach of all, while the excellence of its text will certainly secure for it most favorable commendation —*Cincinnati Lancet and Clinic,* Feb. 16, 1884.

CARPENTER, WM. B., M. D., F. R. S., F. G. S., F. L. S.,
Registrar to the University of London, etc.

Principles of Human Physiology. Edited by HENRY POWER, M. B., Lond., F. R. C. S., Examiner in Natural Sciences, University of Oxford. A new American from the eighth revised and enlarged edition, with notes and additions by FRANCIS G. SMITH, M. D., late Professor of the Institutes of Medicine in the University of Pennsylvania. In one very large and handsome octavo volume of 1083 pages, with two plates and 373 illustrations. Cloth, $5.50; leather, $6.50; half Russia, $7.

SIMON, W., Ph. D., M. D.,
Professor of Chemistry and Toxicology in the College of Physicians and Surgeons, Baltimore, and Professor of Chemistry in the Maryland College of Pharmacy.

Manual of Chemistry. A Guide to Lectures and Laboratory work for Beginners in Chemistry. A Text-book, specially adapted for Students of Pharmacy and Medicine. In one 8vo. vol. of 410 pp., with 16 woodcuts and 7 plates, mostly of actual deposits, with colors illustrating 56 of the most important chemical reactions. Cloth, $3.00; also without plates, cloth, $2.50.

This book supplies a want long felt by students of medicine and pharmacy, and is a concise but thorough treatise on the subject. The long experience of the author as a teacher in schools of medicine and pharmacy is conspicuous in the perfect adaptation of the work to the special needs of the student of these branches. The colored

plates, beautifully executed, illustrating precipitates of various reactions, form a novel and valuable feature of the book, and cannot fail to be appreciated by both student and teacher as a help over the hard places of the science.—*Maryland Medical Journal,* Nov. 22, 1884.

Wöhler's Outlines of Organic Chemistry. Edited by FITTIG. Translated by IRA REMSEN, M. D., Ph. D. In one 12mo. volume of 550 pages. Cloth, $3.

GALLOWAY'S QUALITATIVE ANALYSIS.
LEHMANN'S MANUAL OF CHEMICAL PHYSIOLOGY. In one octavo volume of 327 pages, with 41 illustrations. Cloth, $2.25.

CARPENTER'S PRIZE ESSAY ON THE USE AND ABUSE OF ALCOHOLIC LIQUORS IN HEALTH AND DISEASE. With explanations of scientific words. Small 12mo. 178 pages. Cloth, 60 cents.

FOWNES, GEORGE, Ph. D.

A Manual of Elementary Chemistry; Theoretical and Practical. Embodying WATTS' *Inorganic Chemistry.* New American edition. In one large royal 12mo. volume of 1061 pages, with 168 illustrations on wood and a colored plate. Cloth, $2.75; leather, $3.25.

Fownes' *Chemistry* has been a standard textbook upon chemistry for many years. Its merits are very fully known by chemists and physicians everywhere in this country and in England. As the science has advanced by the making of new discoveries, the work has been revised so as to keep it abreast of the times. It has steadily maintained its position as a text-book with medical students. In this work are treated fully: Heat, Light and Electricity, including Magnetism. The influence exerted by these forces in chemical action upon health and disease, etc., is of the most important kind, and should be familiar to every medical practitioner. We can commend the work as one of the very best text-books upon chemistry extant.—*Cincinnati Medical News,* October, 1885.

Of all the works on chemistry intended for the use of medical students, Fownes' *Chemistry* is perhaps the most widely used. Its popularity is based upon its excellence. This last edition contains all of the material found in the previous, and it is also enriched by the addition of Watts' *Physical and Inorganic Chemistry.* All of the matter is brought to the present standpoint of chemical knowledge. We may safely predict for this work a continuance of the fame and favor it enjoys among medical students.—*New Orleans Medical and Surgical Journal,* March, 1886.

FRANKLAND, E., D. C. L., F.R.S., & JAPP, F. R., F. I. C.,

Professor of Chemistry in the Normal School of Science, London.   *Assist. Prof. of Chemistry in the Normal School of Science, London.*

Inorganic Chemistry. In one handsome octavo volume of 677 pages with 51 woodcuts and 2 lithographic plates. Cloth, $3.75; leather, $4.75.

This work should supersede other works of its class in the medical colleges. It is certainly better adapted than any work upon chemistry, with which we are acquainted, to impart that clear and full knowledge of the science which students of medicine should have. Physicians who feel that their chemical knowledge is behind the times, would do well to devote some of their leisure time to the study of this work. The descriptions and demonstrations are made so plain that there is no difficulty in understanding them.—*Cincinnati Medical News,* January, 1886.

ATTFIELD, JOHN, Ph. D.,

Professor of Practical Chemistry to the Pharmaceutical Society of Great Britain, etc.

Chemistry, General, Medical and Pharmaceutical; Including the Chemistry of the U. S. Pharmacopœia. A Manual of the General Principles of the Science, and their Application to Medicine and Pharmacy. A new American, from the tenth English edition, specially revised by the Author. In one handsome royal 12mo. volume of 728 pages, with 87 illustrations. Cloth, $2.50; leather, $3.00.

A text-book which passes through ten editions in sixteen years must have good qualities. This remark is certainly applicable to Attfield's Chemistry, a book which is so well known that it is hardly necessary to do more than note the appearance of this new and improved edition. It seems, however, desirable to point out that feature of the book which, in all probability, has made it so popular. There can be little doubt that it is its thoroughly practical character, the expression being used in its best sense. The author understands what the student ought to learn, and is able to put himself in the student's place and to appreciate his state of mind.—*American Chemical Journal,* April, 1884.

It is a book on which too much praise cannot be bestowed. As a text-book for medical schools it is unsurpassable in the present state of chemical science, and having been prepared with a special view towards medicine and pharmacy, it is alike indispensable to all persons engaged in those departments of science. It includes the whole chemistry of the last Pharmacopœia.—*Pacific Medical and Surgical Journal,* Jan. 1884.

BLOXAM, CHARLES L.,

Professor of Chemistry in King's College, London.

Chemistry, Inorganic and Organic. New American from the fifth London edition, thoroughly revised and much improved. In one very handsome octavo volume of 727 pages, with 292 illustrations. Cloth, $3.75; leather, $4.75.

Comment from us on this standard work is almost superfluous. It differs widely in scope and aim from that of Attfield, and in its way is equally beyond criticism. It adopts the most direct methods in stating the principles, hypotheses and facts of the science. Its language is so terse and lucid, and its arrangement of matter so logical in sequence that the student never has occasion to complain that chemistry is a hard study. Much attention is paid to experimental illustrations of chemical principles and phenomena, and the mode of conducting these experiments. The book maintains the position it has always held as one of the best manuals of general chemistry in the English language.—*Detroit Lancet,* Feb. 1884.

REMSEN, IRA, M. D., Ph. D.,

Professor of Chemistry in the Johns Hopkins University, Baltimore.

Principles of Theoretical Chemistry, with special reference to the Constitution of Chemical Compounds. Second and revised edition. In one handsome royal 12mo. volume of 240 pages. Cloth, $1.75.

That in so few years a second edition has been called for indicates that many chemical teachers have been found ready to endorse its plan and to adopt its methods. In this edition a considerable proportion of the book has been rewritten, much new matter has been added and the whole has been brought up to date. We earnestly commend this book to every student of chemistry. The high reputation of the author assures its accuracy in all matters of fact, and its judicious conservatism in matters of theory, combined with the fulness with which, in a small compass, the present attitude of chemical science towards the constitution of compounds is considered, gives it a value much beyond that accorded to the average text-books of the day.—*American Journal of Science,* March, 1884.

CHARLES, T. CRANSTOUN, M. D., F. C. S., M. S.,
Formerly Asst. Prof. and Demonst. of Chemistry and Chemical Physics, Queen's College, Belfast.

The Elements of Physiological and Pathological Chemistry. A Handbook for Medical Students and Practitioners. Containing a general account of Nutrition, Foods and Digestion, and the Chemistry of the Tissues, Organs, Secretions and Excretions of the Body in Health and in Disease. Together with the methods for preparing or separating their chief constituents, as also for their examination in detail, and an outline syllabus of a practical course of instruction for students. In one handsome octavo volume of 463 pages, with 38 woodcuts and 1 colored plate. Cloth, $3.50.

The work is thoroughly trustworthy, and informed throughout by a genuine scientific spirit. The author deals with the chemistry of the digestive secretions in a systematic manner, which leaves nothing to be desired, and in reality supplies a want in English literature. The book appears to us to be at once full and systematic, and to show a just appreciation of the relative importance of the various subjects dealt with.—*British Medical Journal*, November 29, 1884.

Dr. Charles' manual admirably fulfils its intention of giving his readers on the one hand a summary, comprehensive but remarkably compact, of the mass of facts in the sciences which have becomne indispensable to the physician; and, on the other hand, of a system of practical directions so minute that analyses often considered formidable may be pursued by any intelligent person.—*Archives of Medicine*, Dec. 1884.

HOFFMANN, F., A.M., Ph.D., & POWER F.B., Ph.D.,
Public Analyst to the State of New York.   Prof. of Anal. Chem. in the Phil. Coll. of Pharmacy.

A Manual of Chemical Analysis, as applied to the Examination of Medicinal Chemicals and their Preparations. Being a Guide for the Determination of their Identity and Quality, and for the Detection of Impurities and Adulterations. For the use of Pharmacists, Physicians, Druggists and Manufacturing Chemists, and Pharmaceutical and Medical Students. Third edition, entirely rewritten and much enlarged. In one very handsome octavo volume of 621 pages, with 179 illustrations. Cloth, $4.25.

We congratulate the author on the appearance of the third edition of this work, published for the first time in this country also. It is admirable and the information it undertakes to supply is both extensive and trustworthy. The selection of processes for determining the purity of the substances of which it treats is excellent and the description of them singularly explicit. Moreover, it is exceptionally free from typographical errors. We have no hesitation in recommending it to those who are engaged either in the manufacture or the testing of medicinal chemicals.—*London Pharmaceutical Journal and Transactions*, 1883.

CLOWES, FRANK, D. Sc., London,
Senior Science-Master at the High School, Newcastle-under-Lyme, etc.

An Elementary Treatise on Practical Chemistry and Qualitative Inorganic Analysis. Specially adapted for use in the Laboratories of Schools and Colleges and by Beginners. Third American from the fourth and revised English edition. In one very handsome royal 12mo. volume of 387 pages, with 55 illustrations. Cloth, $2.50.

The style is clear, the language terse and vigorous. Beginning with a list of apparatus necessary for chemical work, he gradually unfolds the subject from its simpler to its more complex divisions. It is the most readable book of the kind we have yet seen, and is without doubt a systematic, intelligible and fully equipped laboratory guide and text book.—*Medical Record*, July 18, 1885.

We may simply repeat the favorable opinion which we expressed after the examination of the previous edition of this work. It is practical in its aims, and accurate and concise in its statements.—*American Journal of Pharmacy*, August, 1885.

RALFE, CHARLES H., M. D., F. R. C. P.,
Assistant Physician at the London Hospital.

Clinical Chemistry. In one pocket-size 12mo. volume of 314 pages, with 16 illustrations. Limp cloth, red edges, $1.50. See *Students' Series of Manuals*, page 4.

This is one of the most instructive little works that we have met with in a long time. The author is a physician and physiologist, as well as a chemist, consequently the book is unqualifiedly practical, telling the physician just what he ought to know, of the applications of chemistry in medicine. Dr. Ralfe is thoroughly acquainted with the latest contributions to his science, and it is quite refreshing to find the subject dealt with so clearly and simply, yet in such evident harmony with the modern scientific methods and spirit.—*Medical Record*, February 2, 1884.

CLASSEN, ALEXANDER,
Professor in the Royal Polytechnic School, Aix-la-Chapelle.

Elementary Quantitative Analysis. Translated, with notes and additions, by EDGAR F. SMITH, Ph. D., Assistant Professor of Chemistry in the Towne Scientific School, University of Penna. In one 12mo. volume of 324 pages, with 36 illust. Cloth, $2.00.

It is probably the best manual of an elementary nature extant insomuch as its methods are the best. It teaches by examples, commencing with single determinations, followed by separations, and then advancing to the analysis of minerals and such products as are met with in applied chemistry. It is an indispensable book for students in chemistry.—*Boston Journal of Chemistry*, Oct. 1878.

GREENE, WILLIAM H., M. D.,
Demonstrator of Chemistry in the Medical Department of the University of Pennsylvania.

A Manual of Medical Chemistry. For the use of Students. Based upon Bowman's Medical Chemistry. In one 12mo. volume of 310 pages, with 74 illus. Cloth, $1.75.

It is a concise manual of three hundred pages, giving an excellent summary of the best methods of analyzing the liquids and solids of the body, both for the estimation of their normal constituents and the recognition of compounds due to pathological conditions. The detection of poisons is treated with sufficient fulness for the purpose of the student or practitioner.—*Boston Jl. of Chem.*, June,'80.

BRUNTON, T. LAUDER, M.D., D.Sc., F.R.S., F.R.C.P.,
Lecturer on Materia Medica and Therapeutics at St. Bartholomew's Hospital, London, etc.

A Text-book of Pharmacology, Therapeutics and Materia Medica; Including the Pharmacy, the Physiological Action and the Therapeutical Uses of Drugs. In one handsome octavo volume of 1033 pages, with 188 illustrations. Cloth, $5.50; leather, $6.50. *Just ready.*

It is a scientific treatise worthy to be ranked with the highest productions in physiology, either in our own or any other language. Everything is practical, the dry, hard facts of physiology being pressed into service and applied to the treatment of the commonest complaints. The information is so systematically arranged that its availability for immediate use. The index is so carefully compiled that a reference to any special point is at once obtainable. Dr. Brunton is never satisfied with vague generalities, but gives clear and precise directions for prescribing the various drugs and preparations. We congratulate students on being at last placed in possession of a scientific treatise of enormous practical importance.—*The London Lancet*, June 27, 1885.

Dr. Brunton has been building up the material for this volume through sixteen years of steady labor, and the result proves that this long toil was well directed. He has produced a work of singular merit, every page of which is marked by the results of original research, judiciously analyzed. We are not saying too much in pronouncing this treatise the most complete and valuable in our own or any other language on the topic to which it is devoted. The arrangement is eminently scientific. The author is equally satisfactory in all his details, and his work is certainly destined to rank as one of the most important additions to medical literature of the period.—*Medical and Surgical Reporter*, Oct. 17, 1885.

PARRISH, EDWARD,
Late Professor of the Theory and Practice of Pharmacy in the Philadelphia College of Pharmacy.

A Treatise on Pharmacy: designed as a Text-book for the Student, and as a Guide for the Physician and Pharmaceutist. With many Formulæ and Prescriptions. Fifth edition, thoroughly revised, by THOMAS S. WIEGAND, Ph. G. In one handsome octavo volume of 1093 pages, with 256 illustrations. Cloth, $5; leather, $6.

No thoroughgoing pharmacist will fail to possess himself of so useful a guide to practice, and no physician who properly estimates the value of an accurate knowledge of the remedial agents employed by him in daily practice, so far as their miscibility, compatibility and most effective methods of combination are concerned, can afford to leave this work out of the list of their works of reference. The country practitioner, who must always be in a measure his own pharmacist, will find it indispensable.—*Louisville Medical News*, March 29, 1884.

This well-known work presents itself now based upon the recently revised new Pharmacopœia.

Each page bears evidence of the care bestowed upon it, and conveys valuable information from the rich store of the editor's experience. In fact, all that relates to practical pharmacy—apparatus, processes and dispensing—has been arranged and described with clearness in its various aspects, so as to afford aid and advice alike to the student and to the practical pharmacist. The work is judiciously illustrated with good woodcuts.—*American Journal of Pharmacy*, January, 1884.

There is nothing to equal Parrish's *Pharmacy* in this or any other language.—*London Pharmaceutical Journal.*

HERMANN, Dr. L.,
Professor of Physiology in the University of Zurich.

Experimental Pharmacology. A Handbook of Methods for Determining the Physiological Actions of Drugs. Translated, with the Author's permission, and with extensive additions, by ROBERT MEADE SMITH, M. D., Demonstrator of Physiology in the University of Pennsylvania. In one handsome 12mo. volume of 199 pages, with 32 illustrations. Cloth, $1.50.

MAISCH, JOHN M., Phar. D.,
Professor of Materia Medica and Botany in the Philadelphia College of Pharmacy.

A Manual of Organic Materia Medica; Being a Guide to Materia Medica of the Vegetable and Animal Kingdoms. For the use of Students, Druggists, Pharmacists and Physicians. New (second) edition. In one handsome royal 12mo. volume of 526 pages, with 242 illustrations. Cloth, $3.00.

This work contains the substance,—the *practical* "kernel of the nut" picked out, so that the student has no superfluous labor. He can confidently accept what this work places before him, without any fear that the gist of the matter is not in it. Another merit is that the drugs are placed before him in such a manner as to simplify very much the study of them, enabling the mind to grasp them more readily. The illustrations are most excellent, being very true to nature, and are alone worth the price of the book to the student. To the practical physician and pharmacist it is a valuable book for handy reference and for keeping fresh in the memory the knowledge of materia medica and botany already acquired. We can and do heartily recommend it.—*Medical and Surgical Reporter*, Feb. 14, 1885.

BRUCE, J. MITCHELL, M. D., F. R. C. P.,
Physician and Lecturer on Materia Medica and Therapeutics at Charing Cross Hospital, London.

Materia Medica and Therapeutics. An Introduction to Rational Treatment. In one pocket-size 12mo. volume of 555 pages. Limp cloth, $1.50. See *Students' Series of Manuals*, page 4.

GRIFFITH, ROBERT EGLESFIELD, M. D.

A Universal Formulary, containing the Methods of Preparing and Administering Officinal and other Medicines. The whole adapted to Physicians and Pharmaceutists. Third edition, thoroughly revised, with numerous additions, by JOHN M. MAISCH, Phar. D., Professor of Materia Medica and Botany in the Philadelphia College of Pharmacy. In one octavo volume of 775 pages, with 38 illustrations. Cloth, $4.50; leather, $5.50.

STILLÉ, A., M. D., LL. D., & MAISCH, J. M., Phar. D.,
Professor Emeritus of the Theory and Practice of Medicine and of Clinical Medicine in the University of Pennsylvania. *Prof. of Mat. Med. and Botany in Phila. College of Pharmacy, Sec'y to the American Pharmaceutical Association.*

The National Dispensatory: Containing the Natural History, Chemistry, Pharmacy, Actions and Uses of Medicines, including those recognized in the Pharmacopœias of the United States, Great Britain and Germany, with numerous references to the French Codex. Third edition, thoroughly revised and greatly enlarged. In one magnificent imperial octavo volume of 1767 pages, with 311 fine engravings. Cloth, $7.25; leather, $8.00; half Russia, open back, $9.00. With Denison's "Ready Reference Index" $1.00 in addition to price in any of above styles of binding.

In the present revision the authors have labored incessantly with the view of making the third edition of THE NATIONAL DISPENSATORY an even more complete representative of the pharmaceutical and therapeutic science of 1884 than its first edition was of that of 1879. For this, ample material has been afforded not only by the new United States Pharmacopœia, but by those of Germany and France, which have recently appeared and have been incorporated in the Dispensatory, together with a large number of new non-officinal remedies. It is thus rendered the representative of the most advanced state of American, English, French and German pharmacology and therapeutics. The vast amount of new and important material thus introduced may be gathered from the fact that the additions to this edition amount in themselves to the matter of an ordinary full-sized octavo volume, rendering the work larger by twenty-five per cent. than the last edition. The Therapeutic Index (a feature peculiar to this work), so suggestive and convenient to the practitioner, contains 1000 more references than the last edition—the General Index 3700 more, making the total number of references 22,390, while the list of illustrations has been increased by 80. Every effort has been made to prevent undue enlargement of the volume by having in it nothing that could be regarded as superfluous, yet care has been taken that nothing should be omitted which a pharmacist or physician could expect to find in it.

The appearance of the work has been delayed by nearly a year in consequence of the determination of the authors that it should attain as near an approach to absolute accuracy as is humanly possible. With this view an elaborate and laborious series of examinations and tests have been made to verify or correct the statements of the Pharmacopœin, and very numerous corrections have been found necessary. It has thus been rendered indispensable to all who consult the Pharmacopœia.

The work is therefore presented in the full expectation that it will maintain the position universally accorded to it as the standard authority in all matters pertaining to its subject, as registering the furthest advance of the science of the day, and as embodying in a shape for convenient reference the recorded results of human experience in the laboratory, in the dispensing room, and at the bed-side.

Comprehensive in scope, vast in design and splendid in execution, The National Dispensatory may be justly regarded as the most important work of its kind extant.—*Louisville Medical News*, Dec. 6, 1884.

We have much pleasure in recording the appearance of a third edition of this excellent work of reference. It is an admirable abstract of all that relates to chemistry, pharmacy, materia medica, pharmacology and therapeutics. It may be regarded as embodying the Pharmacopœias of the civilized nations of the world, all being brought up to date. The work has been very well done, a large number of extra-pharmacopœial remedies having been added to those mentioned in previous editions.—*London Lancet*, Nov. 22, 1884.

Its completeness as to subjects, the comprehensiveness of its descriptive language, the thoroughness of the treatment of the topics, its brevity not sacrificing the desirable features of information for which such a work is needed, make this volume a marvel of excellence.—*Pharmaceutical Record*, Aug. 15, 1884.

FARQUHARSON, ROBERT, M. D.,
Lecturer on Materia Medica at St. Mary's Hospital Medical School.

A Guide to Therapeutics and Materia Medica. Third American edition, specially revised by the Author. Enlarged and adapted to the U. S. Pharmacopœia by FRANK WOODBURY, M. D. In one handsome 12mo. volume of 524 pages. Cloth, $2.25.

Dr. Farquharson's Therapeutics is constructed upon a plan which brings before the reader all the essential points with reference to the properties of drugs. It impresses these upon him in such a way as to enable him to take a clear view of the actions of medicines and the disordered conditions in which they must prove useful. The double-col-umned pages—one side containing the recognized physiological action of the medicine, and the other the disease in which observers (who are nearly always mentioned) have obtained from it good results—make a very good arrangement. The early chapter containing rules for prescribing is excellent.—*Canada Med. and Surg. Journal*, Dec. 1882.

EDES, ROBERT T., M. D.,
Jackson Professor of Clinical Medicine in Harvard University, Medical Department.

A Text-Book of Materia Medica and Therapeutics. In one octavo volume of about 600 pages, with illustrations. *Preparing.*

STILLÉ, ALFRED, M. D., LL. D.,
Professor of Theory and Practice of Med. and of Clinical Med. in the Univ. of Penna.

Therapeutics and Materia Medica. A Systematic Treatise on the Action and Uses of Medicinal Agents, including their Description and History. Fourth edition, revised and enlarged. In two large and handsome octavo volumes, containing 1936 pages. Cloth, $10.00; leather, $12.00; very handsome half Russia, raised bands, $13.00.

COATS, JOSEPH, M. D., F. F. P. S.,
Pathologist to the Glasgow Western Infirmary.

A Treatise on Pathology. In one very handsome octavo volume of 829 pages, with 339 beautiful illustrations. Cloth, $5.50; leather, $6.50.

The work before us treats the subject of Pathology more extensively than it is usually treated in similar works. Medical students as well as physicians, who desire a work for study or reference, that treats the subjects in the various departments in a very thorough manner, but without prolixity, will certainly give this one the preference to any with which we are acquainted. It sets forth the most recent discoveries, exhibits, in an interesting manner, the changes from a normal condition effected in structures by disease, and points out the characteristics of various morbid agencies, so that they can be easily recognized. But, not limited to morbid anatomy, it explains fully how the functions of organs are disturbed by abnormal conditions. There is nothing belonging to its department of medicine that is not as fully elucidated as our present knowledge will admit.—*Cincinnati Medical News*, Oct. 1883.

GREEN, T. HENRY, M. D.,
Lecturer on Pathology and Morbid Anatomy at Charing-Cross Hospital Medical School, London.

Pathology and Morbid Anatomy. Fifth American from the sixth revised and enlarged English edition. In one very handsome octavo volume of 482 pages, with 150 fine engravings. Cloth, $2.50.

The fact that this well-known treatise has so rapidly reached its sixth edition is a strong evidence of its popularity. The author is to be congratulated upon the thoroughness with which he has prepared this work. It is thoroughly abreast with all the most recent advances in pathology. No work in the English language is so admirably adapted to the wants of the student and practitioner as this, and we would recommend it most earnestly to every one.—*Nashville Journal of Medicine and Surgery*, Nov. 1884.

WOODHEAD, G. SIMS, M. D., F. R. C. P. E.,
Demonstrator of Pathology in the University of Edinburgh.

Practical Pathology. A Manual for Students and Practitioners. In one beautiful octavo volume of 497 pages, with 136 exquisitely colored illustrations. Cloth, $6.00.

It forms a real guide for the student and practitioner who is thoroughly in earnest in his endeavor to see for himself and do for himself. To the laboratory student it will be a helpful companion, and all those who may wish to familiarize themselves with modern methods of examining morbid tissues are strongly urged to provide themselves with this manual. The numerous drawings are not fancied pictures, or merely schematic diagrams, but they represent faithfully the actual images seen under the microscope. The author merits all praise for having produced a valuable work.—*Medical Record*, May 31, 1884.

SCHÄFER, EDWARD A., F. R. S.,
Assistant Professor of Physiology in University College, London.

The Essentials of Histology. In one octavo volume of 246 pages, with 281 illustrations. Cloth, $2.25. *Just ready.*

This admirable work is a cheering example of well-won success, earned by the faithful and diligent pursuit of excellence in presentation of this essential foundation of all true medical science. Since this new work of Professor Schäfer's will doubtless be speedily placed upon the list of text-books required in every medical college, we feel that it needs no further recommendation at our hands.—*Am. Jour. of the Med. Sciences*, Jan. 1886. This short volume might be called a companion book to Green's Pathology, and fills the same place in histology the latter occupies in pathology. This book is so short, clear and satisfactory, as to invite perusal, and repay any time spent in doing so. We think the book deserving of the highest praise.—*New Orleans Medical and Surgical Journal*, Dec. 1885.

CORNIL, V., and RANVIER, L.,
Prof. in the Faculty of Med. of Paris.   *Prof. in the College of France.*

A Manual of Pathological Histology. Translated, with notes and additions, by E. O. SHAKESPEARE, M. D., Pathologist and Ophthalmic Surgeon to Philadelphia Hospital, and by J. HENRY C. SIMES, M. D., Demonstrator of Pathological Histology in the University of Pennsylvania. In one very handsome octavo volume of 800 pages, with 360 illustrations. Cloth, $5.50; leather, $6.50; half Russia, raised bands, $7.

KLEIN, E., M. D., F. R. S.,
Joint Lecturer on General Anat. and Phys. in the Med. School of St. Bartholomew's Hosp., London.

Elements of Histology. In one pocket-size 12mo. volume of 360 pages, with 181 illus. Limp cloth, red edges, $1.50. See *Students' Series of Manuals*, page 4.

Although an elementary work, it is by no means superficial or incomplete, for the author presents in concise language all the fundamental facts regarding the microscopic structure of tissues. The illustrations are numerous and excellent. We commend Dr. Klein's *Elements* most heartily to the student.—*Medical Record*, Dec. 1, 1883.

PEPPER, A. J., M. B., M. S., F. R. C. S.,
Surgeon and Lecturer at St. Mary's Hospital, London.

Surgical Pathology. In one pocket-size 12mo. volume of 511 pages, with 81 illustrations. Limp cloth, red edges, $2.00. See *Students' Series of Manuals*, page 4.

It is not pretentious, but it will serve exceedingly well as a book of reference. It embodies a great deal of matter, extending over the whole field of surgical pathology. Its form is practical, its language is clear, and the information set forth is well-arranged, well-indexed and well-illustrated. The student will find in it nothing that is unnecessary. The list of subjects covers the whole range of surgery. The book supplies a very manifest want and should meet with success.—*New York Medical Journal*, May 31, 1884.

FLINT, AUSTIN, M. D.,
Prof. of the Principles and Practice of Med. and of Clin. Med. in Bellevue Hospital Medical College, N. Y.

A Treatise on the Principles and Practice of Medicine. Designed for the use of Students and Practitioners of Medicine. With an Appendix on the Researches of Koch, and their bearing on the Etiology, Pathology, Diagnosis and Treatment of Phthisis. Fifth edition, revised and largely rewritten In one large and closely-printed octavo volume of 1160 pages. Cloth, $5.50; leather, $6.50; half Russia, $7.

Koch's discovery of the bacillus of tubercle gives promise of being the greatest boon ever conferred by science on humanity, surpassing even vaccination in its benefits to mankind. In the appendix to his work, Professor Flint deals with the subject from a practical standpoint, discussing its bearings on the etiology, pathology, diagnosis, prognosis and treatment of pulmonary phthisis. Thus enlarged and completed, this standard work will be more than ever a necessity to the physician who duly appreciates the responsibility of his calling.

A well-known writer and lecturer on medicine recently expressed an opinion, in the highest degree complimentary of the admirable treatise of Dr. Flint, and in eulogizing it, he described it accurately as "readable and reliable." No text-book is more calculated to enchain the interest of the student, and none better classifies the multitudinous subjects included in it. It has already so far won its way in England, that no inconsiderable number of men use it alone in the study of pure medicine; and we can say of it that it is in every way adapted to serve, not only as a complete guide, but also as an ample instructor in the science and practice of medicine. The style of Dr. Flint is always polished and engaging. The work abounds in perspicuous explanation, and is a most valuable text-book of medicine.—*London Medical News.*

This work is so widely known and accepted as the best American text-book of the practice of medicine that it would seem hardly worth while to give this, the fifth edition, anything more than a passing notice. But even the most cursory examination shows that it is, practically, much more than a revised edition; it is, in fact, rather a new work throughout. This treatise will undoubtedly continue to hold the first place in the estimation of American physicians and students. No one of our medical writers approaches Professor Flint in clearness of diction, breadth of view, and, what we regard of transcendent importance, rational estimate of the value of remedial agents. It is thoroughly *practical*, therefore pre-eminently *the book for American readers.*—*St. Louis Clin. Rec.,* Mar. '81.

HARTSHORNE, HENRY, M. D., LL. D.,
Lately Professor of Hygiene in the University of Pennsylvania.

Essentials of the Principles and Practice of Medicine. A Handbook for Students and Practitioners. Fifth edition, thoroughly revised and rewritten. In one royal 12mo. volume of 669 pages, with 144 illustrations. Cloth, $2.75; half bound, $3.00.

Within the compass of 600 pages it treats of the history of medicine, general pathology, general symptomatology, and physical diagnosis (including laryngoscope, ophthalmoscope, etc.), general therapeutics, nosology, and special pathology and practice. There is a wonderful amount of information contained in this work, and it is one of the best of its kind that we have seen.—*Glasgow Medical Journal,* Nov. 1882.

An indispensable book. No work ever exhibited a better average of actual practical treatment than this one; and probably not one writer in our day had a better opportunity than Dr. Hartshorne for condensing all the views of eminent practitioners into a 12mo. The numerous illustrations will be very useful to students especially. These essentials, as the name suggests, are not intended to supersede the text-books of Flint and Bartholow, but they are the most valuable in affording the means to see at a glance the whole literature of any disease, and the most valuable treatment.—*Chicago Medical Journal and Examiner,* April, 1882.

BRISTOWE, JOHN SYER, M. D., F. R. C. P.,
Physician and Joint Lecturer on Medicine at St. Thomas' Hospital, London.

A Treatise on the Practice of Medicine. Second American edition, revised by the Author. Edited, with additions, by JAMES H. HUTCHINSON, M.D., physician to the Pennsylvania Hospital. In one handsome octavo volume of 1085 pages, with illustrations. Cloth, $5.00; leather, $6.00; very handsome half Russia, raised bands, $6.50.

The reader will find every conceivable subject connected with the practice of medicine ably presented, in a style at once clear, interesting and concise. The additions made by Dr. Hutchinson are appropriate and practical, and greatly add to its usefulness to American readers.—*Buffalo Medical and Surgical Journal,* March, 1880.

WATSON, SIR THOMAS, M. D.,
Late Physician in Ordinary to the Queen.

Lectures on the Principles and Practice of Physic. A new American from the fifth English edition. Edited, with additions, and 190 illustrations, by HENRY HARTSHORNE, A. M., M. D., late Professor of Hygiene in the University of Pennsylvania. In two large octavo volumes of 1840 pages. Cloth, $9.00; leather, $11.00.

LECTURES ON THE STUDY OF FEVER. By A. HUDSON, M. D., M. R. I. A. In one octavo volume of 308 pages. Cloth, $2.50.
STOKES' LECTURES ON FEVER. Edited by John William Moore, M. D., F. K. Q. C. P. In one octavo volume of 280 pages. Cloth, $2.00.

A TREATISE ON FEVER. By ROBERT D. LYONS, K. C. C. In one 8vo. vol. of 354 pp. Cloth, $2.25.
LA ROCHE ON YELLOW FEVER, considered in its Historical, Pathological, Etiological and Therapeutical Relations. In two large and handsome octavo volumes of 1468 pp. Cloth, $7.00.

A CENTURY OF AMERICAN MEDICINE, 1776–1876. By Drs. E. H. CLARKE, H. J. BIGELOW, S. D. GROSS, T. G. THOMAS, and J. S. BILLINGS. In one 12mo. volume of 370 pages. Cloth, $2.25.

For Sale by Subscription Only.

A System of Practical Medicine.
BY AMERICAN AUTHORS.
EDITED BY WILLIAM PEPPER, M. D., LL. D.,
PROVOST AND PROFESSOR OF THE THEORY AND PRACTICE OF MEDICINE AND OF CLINICAL MEDICINE IN THE UNIVERSITY OF PENNSYLVANIA,
Assisted by LOUIS STARR, M. D., Clinical Professor of the Diseases of Children in the Hospital of the University of Pennsylvania.

In five imperial octavo volumes, containing about 1100 pages each, with illustrations. Price per volume, cloth, $5; leather, $6; half Russia, raised bands and open back, $7. Volumes I., II., III. and IV., containing 4315 pages and 140 illustrations, are now ready. Volume V. will be ready in June.

In this great work American medicine will be for the first time represented by its worthiest teachers, and presented in the full development of the practical utility which is its preëminent characteristic. The most able men—from the East and the West, from the North and the South, from all the prominent centres of education, and from all the hospitals which afford special opportunities for study and practice—have united in generous rivalry to bring together this vast aggregate of specialized experience.

The distinguished editor has so apportioned the work that each author has had assigned to him the subject which he is peculiarly fitted to discuss, and in which his views will be accepted as the latest expression of scientific and practical knowledge. The practitioner will therefore find these volumes a complete, authoritative and unfailing work of reference, to which he may at all times turn with full certainty of finding what he needs in its most recent aspect, whether he seeks information on the general principles of medicine, or minute guidance in the treatment of special disease. So wide is the scope of the work that, with the exception of midwifery and matters strictly surgical, it embraces the whole domain of medicine, including the departments for which the physician is accustomed to rely on special treatises, such as diseases of women and children, of the genito-urinary organs, of the skin, of the nerves, hygiene and sanitary science, and medical ophthalmology and otology. Moreover, authors have inserted the formulas which they have found most efficient in the treatment of the various affections. It may thus be truly regarded as a COMPLETE LIBRARY OF PRACTICAL MEDICINE, and the general practitioner possessing it may feel secure that he will require little else in the daily round of professional duties.

In spite of every effort to condense the vast amount of practical information furnished, it has been impossible to present it in less than 5 large octavo volumes, containing about 5500 beautifully printed pages, and embodying the matter of about 15 ordinary octavos. Illustrations are introduced wherever requisite to elucidate the text.

As the complete material for the work is in the hands of the editor, the profession may confidently await the appearance of the remaining volumes upon the dates above specified. A detailed prospectus of the work will be sent to any address on application to the publishers.

This magnificent work has filled us with feelings of warm admiration. It is adorned with a galaxy of famous names, many of them familiar to the European names as representative of the best work done in scientific medicine in the Western Continent, and the articles are therefore to be regarded as coming from the highest authorities on the particular subjects of which they treat. We would offer our congratulations on the excellence of the *System of Medicine*, and in expressing our high opinion of the work we have only to add our hearty wish that it may be read as much in this country as it deserves.—*Edinburgh Medical Journal*, February, 1886.

The third volume of this great work, which attained a merited popularity immediately on the issue of the first volume, is in no way inferior to its predecessors. This "System of Medicine by American Authors" is a monument to American medicine.—*Journal of American Medical Association*, December 5, 1886.

We consider it one of the grandest works in Practical Medicine in the English language. It is a work of which the profession of this country can feel proud. Written exclusively by American physicians who are acquainted with all the varieties of climate in the United States, the character of the soil, the manners and customs of the people, etc., it is peculiarly adapted to the wants of American practitioners of medicine, and it seems to us that every one of them would desire to have it. It has been truly called a "Complete Library of Practical Medicine," and the general practitioner will require little else in his round of professional duties.—*Cincinnati Medical News*, March, 1886.

REYNOLDS, J. RUSSELL, M. D.,
Professor of the Principles and Practice of Medicine in University College, London.

A System of Medicine. With notes and additions by HENRY HARTSHORNE, A. M., M. D., late Professor of Hygiene in the University of Pennsylvania. In three large and handsome octavo volumes, containing 3056 double-columned pages, with 317 illustrations. Price per volume, cloth, $5.00; sheep, $6.00; very handsome half Russia, raised bands, $6.50. Per set, cloth, $15; leather, $18; half Russia, $19.50. *Sold only by subscription.*

16   LEA BROTHERS & CO.'S PUBLICATIONS—Clinical Med., etc.

FOTHERGILL, J. M., M. D., Edin., M. R. C. P., Lond.,
Physician to the City of London Hospital for Diseases of the Chest.

The Practitioner's Handbook of Treatment; Or, The Principles of Therapeutics. New edition. In one octavo volume. *Shortly.*

From the Preface to the Previous Edition.

This work is not an imperfect Practice of Physic, but an attempt of original character to explain the *rationale* of our therapeutic measures. First the physiology of each subject is given, then the pathology is reviewed, so far as they bear on the treatment; next the action of remedies is examined; after which their practical application in concrete prescriptions is furnished. It is designed to furnish to the practitioner reasons for the faith which is in him; and is a work on medical tactics for the bedside rather than the examination table.

STILLÉ, ALFRED, M. D., LL. D.,
Professor Emeritus of the Theory and Practice of Med. and of Clinical Med. in the Univ. of Penna.

Cholera: Its Origin, History, Causation, Symptoms, Lesions, Prevention and Treatment. In one handsome 12mo. volume of 163 pages, with a chart. Cloth, $1.25. *Just ready.*

This timely little work is full of the learning and good judgment which marks all that comes from the pen of its distinguished author. What he has to say on treatment is characterized by his usual caution and his well-known preference | for a rational system. Altogether, the monograph is one that will have an excellent influence on the professional mind.—*Medical and Surgical Reporter*, August 1, 1885. q.

FLINT, AUSTIN, M. D.

Clinical Medicine. A Systematic Treatise on the Diagnosis and Treatment of Diseases. Designed for Students and Practitioners of Medicine. In one large and handsome octavo volume of 799 pages. Cloth, $4.50; leather, $5.50; half Russia, $6.00.

It is here that the skill and learning of the great clinician are displayed. He has given us a storehouse of medical knowledge, excellent for the student, convenient for the practitioner, the result of a long life of the most faithful clinical work, collected by an energy as vigilant and systematic as untiring, and weighed by a judgment no less clear than his observation is close.—*Archives of Medicine*, Dec. 1879.

To give an adequate and useful conspectus of the extensive field of modern clinical medicine is a task of no ordinary difficulty; but to accomplish this consistently with brevity and clearness, the different subjects and their several parts receiving the attention which, relatively to their importance, medical opinion claims for them, is still more difficult. This task, we feel bound to say, has been executed with more than partial success by Dr. Flint, whose name is already familiar to students of advanced medicine in this country as that of the author of two works of great merit on special subjects, and of numerous papers exhibiting much originality and extensive research.—*The Dublin Journal*, Dec. 1879.

By the Same Author.

Essays on Conservative Medicine and Kindred Topics. In one very handsome royal 12mo. volume of 210 pages. Cloth, $1.38.

FINLAYSON, JAMES, M. D., Editor,
Physician and Lecturer on Clinical Medicine in the Glasgow Western Infirmary, etc.

Clinical Diagnosis. A Handbook for Students and Practitioners of Medicine. With Chapters by Prof. Gairdner on the Physiognomy of Disease; Prof. Stephens on Diseases of the Female Organs; Dr. Robertson on Insanity; Dr. Gemmell on Physical Diagnosis; Dr. Coats on Laryngoscopy and Post-Mortem Examinations, and by the Editor on Case-taking, Family History and Symptoms of Disorder in the Various Systems. New edition. In one handsome 12mo. volume of 600 pages, with about 100 illustrations. *Preparing.*

BROADBENT, W. H., M. D., F. R. C. P.,
Physician to and Lecturer on Medicine at St. Mary's Hospital.

The Pulse. In one 12mo. volume. *Preparing.* See *Series of Clinical Manuals*, page 4.

FENWICK, SAMUEL, M. D.,
Assistant Physician to the London Hospital.

The Student's Guide to Medical Diagnosis. From the third revised and enlarged English edition. In one very handsome royal 12mo. volume of 328 pages, with 87 illustrations on wood. Cloth, $2.25.

TANNER, THOMAS HAWKES, M. D.

A Manual of Clinical Medicine and Physical Diagnosis. Third American from the second London edition. Revised and enlarged by TILBURY FOX, M. D. In one small 12mo. volume of 362 pages, with illustrations. Cloth, $1.50.

STURGES' INTRODUCTION TO THE STUDY OF CLINICAL MEDICINE. Being a Guide to the Investigation of Disease. In one handsome 12mo. volume of 127 pages. Cloth, $1.25. | DAVIS' CLINICAL LECTURES ON VARIOUS IMPORTANT DISEASES. By N. S. DAVIS, M. D. Edited by FRANK H. DAVIS, M. D. Second edition. 12mo. 287 pages. Cloth, $1.75.

RICHARDSON, B. W., M.A., M.D., LL. D., F.R.S., F.S.A.
Fellow of the Royal College of Physicians, London.

Preventive Medicine. In one octavo volume of 729 pages. Cloth, $4; leather, $5; very handsome half Russia, raised bands, $5.50.

Dr. Richardson has succeeded in producing a work which is elevated in conception, comprehensive in scope, scientific in character, systematic in arrangement, and which is written in a clear, concise and pleasant manner. He evinces the happy faculty of extracting the pith of what is known on the subject, and of presenting it in a most simple, intelligent and practical form. There is perhaps no similar work written for the general public that contains such a complete, reliable and instructive collection of data upon the diseases common to the race, their origins, causes, and the measures for their prevention. The descriptions of diseases are clear, chaste and scholarly; the discussion of the question of disease is comprehensive, masterly and fully abreast with the latest and best knowledge on the subject, and the preventive measures advised are accurate, explicit and reliable.—*The American Journal of the Medical Sciences*, April, 1884.

This is a book that will surely find a place on the table of every progressive physician. To the medical profession, whose duty is quite as much to prevent as to cure disease, the book will be a boon.—*Boston Medical and Surgical Journal*, Mar. 6, 1884.

The treatise contains a vast amount of solid, valuable hygienic information.—*Medical and Surgical Reporter*, Feb. 23, 1884.

BARTHOLOW, ROBERTS, A. M., M. D., LL. D.,
Prof. of Materia Medica and General Therapeutics in the Jefferson Med. Coll. of Phila., etc.

Medical Electricity. A Practical Treatise on the Applications of Electricity to Medicine and Surgery. New (third) edition. In one very handsome octavo volume of 300 pages, with about 125 illustrations. *Preparing.*
A notice of the previous edition is appended.

A most excellent work, addressed by a practitioner to his fellow-practitioners, and therefore thoroughly practical. The work now before us has the exceptional merit of clearly pointing out where the benefits to be derived from electricity must come. It contains all and everything that the practitioner needs in order to understand intelligently the nature and laws of the agent he is making use of, and for its proper application in practice. In a condensed, practical form, it presents to the physician all that he would wish to remember after perusing a whole library on medical electricity, including the results of the latest investigations. It is the book for the practitioner, and the necessity for a second edition proves that it has been appreciated by the profession.—*Physician and Surgeon*, Dec. 1882.

THE YEAR-BOOK OF TREATMENT FOR 1885.
A Comprehensive and Critical Review for Practitioners of Medicine. In one 12mo. volume of 320 pages, bound in limp cloth, $1.25. *Just ready.*

One strong feature of the book is that treatment comes in for a greater share of attention than pathology, and this gives to it a practical nature—because, what the practitioner wants to know is not a theory or a scientific explanation, but what is the best thing for him to do in certain emergencies, or how he can improve his treatment. He can learn by consulting the *Year-Book of Treatment* what has been done all over the world by the best practitioners in medicine and surgery, in every department and in every specialty. As the book is arranged in sections, there can be little difficulty in finding out what may be required; and as the descriptions of the newer methods of operation are very full, even new operations can be tried.—*Provincial Medical Journal*, March 1, '86.

THE YEAR-BOOK OF TREATMENT FOR 1884.
Similar to that of 1885 above.

It is a complete account of the more important advances made in the treatment of disease. Extreme pains have been taken to explain clearly in the fewest possible words the views of each writer, and the details of each subject. One of the principle points about the book is its practical, yet concise language. Each editor has well performed his duty, and we can say with truth that it is a volume well worth buying for frequent use.—*Virginia Medical Monthly*, March, 1885.

In a few moments the busy practitioner can refresh his mind as to the principal advances in treatment for a year past. This kind of work is peculiarly useful at the present time, when current literature is teeming with innumerable so-called advances, of which the practitioner has not time to determine the value. Here he has, collected from many sources, a *résumé* of the theories and facts which are new, either entirely or in part, the decision as to their novelty being made by those who by wide reading and long experience are fully competent to render such a verdict.—*American Journal of the Medical Sciences*, April, 1885.

HABERSHON, S. O., M. D.,
Senior Physician to and late Lect. on Principles and Practice of Med. at Guy's Hospital, London.

On the Diseases of the Abdomen; Comprising those of the Stomach, and other parts of the Alimentary Canal, Œsophagus, Cæcum, Intestines and Peritoneum. Second American from third enlarged and revised English edition. In one handsome octavo volume of 554 pages, with illustrations. Cloth, $3.50.

SCHREIBER, DR. JOSEPH.
A Manual of Treatment by Massage and Methodical Muscle Exercise. Translated by WALTER MENDELSON, M. D., of New York. In one handsome octavo volume of about 300 pages, with about 125 fine engravings. *Preparing.*

TODD'S CLINICAL LECTURES ON CERTAIN ACUTE DISEASES. In one octavo volume of 320 pages. Cloth, $2.50.

HOLLAND'S MEDICAL NOTES AND REFLECTIONS. 1 vol. 8vo.; pp. 493. Cloth, $3.50.

FLINT, AUSTIN, M. D.,
Professor of the Principles and Practice of Medicine in Bellevue Hospital Medical College, N. Y.

A Manual of Auscultation and Percussion; Of the Physical Diagnosis of Diseases of the Lungs and Heart, and of Thoracic Aneurism. Fourth edition. In one handsome royal 12mo. volume of 278 pages, with 14 illustrations. Cloth, $1.75. *Just ready.*

The student needs a first-class text-book which the subject is fully explained for him to study, Dr. Flint's work is just such a book. It contains the substance of the lessons which the author has for many years given in connection with practical instruction in auscultation and percussion to private classes, composed of medical students and practitioners. The fact that within two years a large edition of this manual has been exhausted, is proof of the favor with which it has been regarded by the medical profession.—*Cincinnati Medical News*, Feb. 1886.

BY THE SAME AUTHOR.

Physical Exploration of the Lungs by Means of Auscultation and Percussion. Three lectures delivered before the Philadelphia County Medical Society, 1882-83. In one handsome small 12mo. volume of 83 pages. Cloth, $1.00.

A Practical Treatise on the Physical Exploration of the Chest and the Diagnosis of Diseases Affecting the Respiratory Organs. Second and revised edition. In one handsome octavo volume of 591 pages. Cloth, $4.50.

Phthisis: Its Morbid Anatomy, Etiology, Symptomatic Events and Complications, Fatality and Prognosis, Treatment and Physical Diagnosis; In a series of Clinical Studies. In one handsome octavo volume of 442 pages. Cloth, $3.50.

A Practical Treatise on the Diagnosis, Pathology and Treatment of Diseases of the Heart. Second revised and enlarged edition. In one octavo volume of 550 pages, with a plate. Cloth, $4.

COHEN, J. SOLIS, M. D.,
Lecturer on Laryngoscopy and Diseases of the Throat and Chest in the Jefferson Medical College.

Diseases of the Throat and Nasal Passages. A Guide to the Diagnosis and Treatment of Affections of the Pharynx, Œsophagus, Trachea, Larynx and Nares. Third edition, thoroughly revised and rewritten, with a large number of new illustrations. In one very handsome octavo volume. *Preparing.*

SEILER, CARL, M. D.,
Lecturer on Laryngoscopy in the University of Pennsylvania.

A Handbook of Diagnosis and Treatment of Diseases of the Throat, Nose and Naso-Pharynx. Second edition. In one handsome royal 12mo. volume of 294 pages, with 77 illustrations. Cloth, $1.75.

It is one of the best of the practical text-books on this subject with which we are acquainted. The present edition has been increased in size, but its eminently practical character has been maintained. Many new illustrations have also been introduced, a case-record sheet has been added, and there are a valuable bibliography and a good index of the whole. For any one who wishes to make himself familiar with the practical management of cases of throat and nose disease, the book will be found of great value.—*New York Medical Journal*, June 9, 1883.

BROWNE, LENNOX, F. R. C. S., Edin.,
Senior Surgeon to the Central London Throat and Ear Hospital, etc.

The Throat and its Diseases. Second American from the second English edition, thoroughly revised. With 100 typical illustrations in colors and 50 wood engravings, designed and executed by the Author. In one very handsome imperial octavo volume of about 350 pages. *Preparing.*

GROSS, S. D., M.D., LL.D., D.C.L. Oxon., LL.D. Cantab.

A Practical Treatise on Foreign Bodies in the Air-passages. In one octavo volume of 452 pages, with 59 illustrations. Cloth, $2.75.

FULLER ON DISEASES OF THE LUNGS AND AIR-PASSAGES. Their Pathology, Physical Diagnosis, Symptoms and Treatment. From the second and revised English edition. In one octavo volume of 475 pages. Cloth, $3.50.

SLADE ON DIPHTHERIA; Its Nature and Treatment, with an account of the History of its Prevalence in various Countries. Second and revised edition. In one 12mo. vol., pp. 158. Cloth, $1.25.

WALSHE ON THE DISEASES OF THE HEART AND GREAT VESSELS. Third American edition. In 1 vol. 8vo., 416 pp. Cloth, $3.00.

PAVY'S TREATISE ON THE FUNCTION OF DIGESTION; its Disorders and their Treatment. From the second London edition. In one octavo volume of 238 pages. Cloth, $2.00.

CHAMBERS' MANUAL OF DIET AND REGIMEN IN HEALTH AND SICKNESS. In one handsome octavo volume of 302 pp. Cloth, $2.75.

SMITH ON CONSUMPTION; its Early and Remediable Stages. 1 vol. 8vo., pp. 253. Cloth, $2.25.

LA ROCHE ON PNEUMONIA. 1 vol. 8vo. of 490 pages. Cloth, $3.00.

WILLIAMS ON PULMONARY CONSUMPTION; its Nature, Varieties and Treatment. With an analysis of one thousand cases to exemplify its duration. In one 8vo. vol. of 303 pp. Cloth, $2.50.

JONES' CLINICAL OBSERVATIONS ON FUNCTIONAL NERVOUS DISORDERS. Second American edition. In one handsome octavo volume of 340 pages. Cloth, $3.25.

BARLOW'S MANUAL OF THE PRACTICE OF MEDICINE. With additions by D. F. CONDIE, M. D. 1 vol. 8vo., pp. 603. Cloth, $2.50.

ROSS, JAMES, M. D., F. R. C. P., LL. D.,
Senior Assistant Physician to the Manchester Royal Infirmary.

A Handbook on Diseases of the Nervous System. In one octavo volume of 725 pages, with 184 illustrations. Cloth, $4.50; leather, $5.50. *Just ready.*

Dr. Ross' reputation as a neurologist is so well established that anything we can say will scarcely add to it. This work is a condensation of his large treatise, intended for the use of students and for the "busy practitioner." As coming from Dr. Ross' pen the work can scarcely be less than authoritative. It is besides, clear, succinct and readable. The descriptions are always graphic—sometimes almost photographic. The treatment is always as definite as circumstances will permit. The work is altogether good—too good we fear for the success of his larger work.—*The Bristol Medico-Chirurgical Journal*, March, 1886.

MITCHELL, S. WEIR, M. D.,
Physician to Orthopædic Hospital and the Infirmary for Diseases of the Nervous System, Phila., etc.

Lectures on Diseases of the Nervous System; Especially in Women. Second edition. In one 12mo. volume of 288 pages. Cloth, $1.75.

No work in our language develops or displays more features of that many-sided affection, hysteria, or gives clearer directions for its differentiation, or sounder suggestions relative to its general management and treatment. The book is particularly valuable in that it represents in which the main the author's own clinical studies, which have been so extensive and fruitful as to give his teachings the stamp of authority all over the realm of medicine. The work, although written by a specialist, has no exclusive character, and the general practitioner above all others will find its perusal profitable, since it deals with diseases he frequently encounters and must essay to treat.—*American Practitioner*, August, 1885.

HAMILTON, ALLAN McLANE, M. D.,
Attending Physician at the Hospital for Epileptics and Paralytics, Blackwell's Island, N. Y.

Nervous Diseases; Their Description and Treatment. Second edition, thoroughly revised and rewritten. In one octavo volume of 598 pages, with 72 illustrations. Cloth, $4.

When the first edition of this good book appeared we gave it our emphatic endorsement, and the present edition enhances our appreciation of the book and its author as a safe guide to students of clinical neurology. One of the best and most critical of English neurological journals, *Brain*, has characterized this book as the best of its kind in any language, which is a handsome endorsement from an exalted source. The improvements in the new edition, and the additions to it, will justify its purchase even by those who possess the old.—*Alienist and Neurologist*, April, 1882.

TUKE, DANIEL HACK, M. D.,
Joint Author of The Manual of Psychological Medicine, etc.

Illustrations of the Influence of the Mind upon the Body in Health and Disease. Designed to elucidate the Action of the Imagination. New edition. Thoroughly revised and rewritten. In one handsome octavo volume of 467 pages, with two colored plates. Cloth, $3.00.

It is impossible to peruse these interesting chapters without being convinced of the author's perfect sincerity, impartiality, and thorough mental grasp. Dr. Tuke has exhibited the requisite amount of scientific address on all occasions, and the more intricate the phenomena the more firmly has he adhered to a physiological and rational method of interpretation. Guided by an enlightened deduction, the author has reclaimed for science a most interesting domain in psychology, previously abandoned to charlatans and empirics. This book, well conceived and well written, must commend itself to every thoughtful understanding.—*New York Medical Journal*, September 6, 1884.

CLOUSTON, THOMAS S., M. D., F. R. C. P., L. R. C. S.,
Lecturer on Mental Diseases in the University of Edinburgh.

Clinical Lectures on Mental Diseases. With an Appendix, containing an Abstract of the Statutes of the United States and of the Several States and Territories relating to the Custody of the Insane. By CHARLES F. FOLSOM, M. D., Assistant Professor of Mental Diseases, Med. Dep. of Harvard Univ. In one handsome octavo volume of 541 pages, with eight lithographic plates, four of which are beautifully colored. Cloth, $4.

The practitioner as well as the student will accept the plain, practical teaching of the author as a forward step in the literature of insanity. It is refreshing to find a physician of Dr. Clouston's experience and high reputation giving the bedside notes upon which his experience has been founded and his mature judgment established. Such clinical observations cannot but be useful to the general practitioner in guiding him to a diagnosis and indicating the treatment, especially in many obscure and doubtful cases of mental disease. To the American reader Dr. Folsom's *Appendix* adds greatly to the value of the work, and will make it a desirable addition to every library. —*American Psychological Journal*, July, 1884.

☞ Dr. Folsom's *Abstract* may also be obtained separately in one octavo volume of 108 pages. Cloth, $1.50.

SAVAGE, GEORGE H., M. D.,
Lecturer on Mental Diseases at Guy's Hospital, London.

Insanity and Allied Neuroses, Practical and Clinical. In one 12mo. vol. of 551 pages, with 18 illus. Cloth, $2.00. See *Series of Clinical Manuals*, page 4.

PLAYFAIR, W. S., M. D., F. R. C. P.,

The Systematic Treatment of Nerve Prostration and Hysteria. In one handsome small 12mo. volume of 97 pages. Cloth, $1.00.

Blandford on Insanity and its Treatment: Lectures on the Treatment, Medical and Legal, of Insane Patients. In one very handsome octavo volume.

ASHHURST, JOHN, Jr., M. D.,
Professor of Clinical Surgery, Univ. of Penna., Surgeon to the Episcopal Hospital, Philadelphia.

The Principles and Practice of Surgery. New (fourth) edition, enlarged and revised. In one large and handsome octavo volume of 1114 pages, with 597 illustrations. Cloth, $6; leather, $7; half Russia, $7.50. *Just ready.*

As with Erichsen so with Ashhurst, its position in professional favor is established, and one has now but to notice the changes, if any, in theory and practice, that are apparent in the present as compared with the preceding edition, published three years ago. The work has been brought well up to date, and is larger and better illustrated than before, and its author may rest assured that it will certainly have a "continuance of the favor with which it has heretofore been received."—*The American Journal of the Medical Sciences,* Jan. 1886.

Every advance in surgery worth notice, chronicled in recent literature, has been suitably recognized and noted in its proper place. Suffice it to say, we regard Ashhurst's Surgery, as now presented in the fourth edition, as the best single volume on surgery published in the English language, valuable alike to the student and the practitioner, to the one as a text-book, to the other as a manual of practical surgery. With pleasure we give this volume our endorsement in full.—*New Orleans Medical and Surgical Journal,* Jan., 1886.

GROSS, S. D., M. D., LL. D., D. C. L. Oxon., LL. D. Cantab.,
Emeritus Professor of Surgery in the Jefferson Medical College of Philadelphia.

A System of Surgery: Pathological, Diagnostic, Therapeutic and Operative. Sixth edition, thoroughly revised and greatly improved. In two large and beautifully-printed imperial octavo volumes containing 2382 pages, illustrated by 1623 engravings. Strongly bound in leather, raised bands, $15; half Russia, raised bands, $16.

Dr. Gross' *System of Surgery* has long been the standard work on that subject for students and practitioners.—*London Lancet,* May 10, 1884.

The work as a whole needs no commendation. Many years ago it earned for itself the enviable reputation of the leading American work on surgery, and it is still capable of maintaining that standard. A considerable amount of new material has been introduced, and altogether the distinguished author has reason to be satisfied that he has placed the work fully abreast of the state of our knowledge.—*Med. Record,* Nov. 18, 1882.

His *System of Surgery,* which, since its first edition in 1859, has been a standard work in this country as well as in America, in the whole domain of surgery," tells how earnest and laborious and wise a surgeon he was, how thoroughly he appreciated the work done by men in other countries, and how much he contributed to promote the science and practice of surgery in his own. There has been no man to whom America is so much indebted in this respect as the Nestor of surgery.—*British Medical Journal,* May 10, 1884.

GOULD, A. PEARCE, M. S., M. B., F. R. C. S.,
Assistant Surgeon to Middlesex Hospital.

Elements of Surgical Diagnosis. In one pocket-size 12mo. volume of 589 pages. Cloth, $2.00. See *Students' Series of Manuals,* page 4.

This book will be found to be a most useful guide for the hard-worked practitioner. Mr. Gould's style is eminently clear and precise, and we can cordially recommend the manual as being the outcome of the efforts of an honest and thoroughly practical surgeon.—*The Medical News,* Jan. 24, 1885.

This is a capital little book, written by a practical man on a very practical subject. The topics are very systematically and succinctly arranged, are tersely presented, and the points of diagnosis very intelligently discussed. It will be found to be of the greatest amount of help both to teacher and student.—*Medical Record,* Feb. 28, 1885.

GIBNEY, V. P., M. D.,
Surgeon to the Orthopædic Hospital, New York, etc.

Orthopædic Surgery. For the use of Practitioners and Students. In one handsome octavo volume, profusely illustrated. *Preparing.*

DRUITT, ROBERT, M. R. C. S., etc.

The Principles and Practice of Modern Surgery. From the eighth London edition. In one 8vo. volume of 687 pages, with 432 illus. Cloth, $4; leather, $5.

ROBERTS, JOHN B., A. M., M. D.,
Lecturer on Anatomy and on Operative Surgery at the Philadelphia School of Anatomy.

The Principles and Practice of Modern Surgery. For the use of Students and Practitioners of Medicine and Surgery. In one very handsome octavo volume of about 500 pages, with many illustrations. *Preparing.*

BELLAMY, EDWARD, F. R. C. S.,
Surgeon and Lecturer on Surgery at Charing Cross Hospital, London.

Operative Surgery. *Shortly.* See *Students' Series of Manuals,* page 4.

SARGENT ON BANDAGING AND OTHER OPERATIONS OF MINOR SURGERY. New edition, with a Chapter on military surgery. One 12mo. volume of 383 pages, with 187 cuts. Cloth, $1.75.

PIRRIE'S PRINCIPLES AND PRACTICE OF SURGERY. Edited by JOHN NEILL, M. D. In one 8vo. vol. of 784 pp. with 316 illus. Cloth, $3.75.

SKEY'S OPERATIVE SURGERY. In one vol. 8vo. of 661 pages, with 81 woodcuts. Cloth, $3.25.

MILLER'S PRINCIPLES OF SURGERY. Fourth American from the third Edinburgh edition. In one 8vo. vol. of 638 pages, with 340 illustrations. Cloth, $3.75.

MILLER'S PRACTICE OF SURGERY. Fourth and revised American from the last Edinburgh edition. In one large 8vo. vol. of 682 pages, with 364 illustrations. Cloth, $3.75.

ERICHSEN, JOHN E., F. R. S., F. R. C. S.,
Professor of Surgery in University College, London, etc.

The Science and Art of Surgery; Being a Treatise on Surgical Injuries, Diseases and Operations. From the eighth and enlarged English edition. In two large and beautiful octavo volumes of 2316 pages, illustrated with 984 engravings on wood. Cloth, $9; leather, raised bands, $11; half Russia, raised bands, $12.

In noticing the eighth edition of this well-known work, it would appear superfluous to say more than that it has, like its predecessors, been brought fully up to the times, and is in consequence one of the best treatises upon surgery that has ever been penned by one man. We have always regarded "The Science and Art of Surgery" as one of the best surgical text-books in the English language, and this eighth edition only confirms our previous opinion. We take great pleasure in cordially commending it to our readers.—*The Medical News*, April 11, 1885.

After being before the profession for thirty years and maintaining during that period a reputation as a leading work on surgery, there is not much to be said in the way of comment or criticism. That it still holds its own goes without saying. The author infuses into it his large experience and ripe judgment. Wedded to no school, committed to no theory, biassed by no hobby, he imparts an honest personality in his observations, and his teachings are the rulings of an impartial judge. Such men are always safe guides, and their works stand the tests of time and experience. Such an author is Erichsen, and such a work is his *Surgery.—Medical Record*, Feb. 21, 1885.

BRYANT, THOMAS, F. R. C. S.,
Surgeon and Lecturer on Surgery at Guy's Hospital, London.

The Practice of Surgery. Fourth American from the fourth and revised English edition. In one large and very handsome imperial octavo volume of 1040 pages, with 727 illustrations. Cloth, $6.50; leather, $7.50; half Russia, $8.00.

The treatise takes in the whole field of surgery, that of the eye, the ear, the female organs, orthopædics, venereal diseases, and military surgery, as well as more common and general topics. All of these are treated with clearness and with sufficient fulness to suit all practical purposes. The illustrations are numerous and well printed. We do not doubt that this new edition will continue to maintain the popularity of this standard work.—*Medical and Surgical Reporter*, Feb. 14, '85.

This most magnificent work upon surgery has reached a fourth edition in this country, showing the high appreciation in which it is held by the American profession. It comes fresh from the pen of the author. That it is the very best work on surgery for medical students we think there can be no doubt. The author seems to have understood just what a student needs, and has prepared the work accordingly.—*Cincinnati Medical News*, January, 1885.

By the same Author.
Diseases of the Breast. In one 12mo. volume. *Preparing.* See *Series of Clinical Manuals,* page 4.

TREVES, FREDERICK, F. R. C. S.,
Hunterian Professor at the Royal College of Surgeons of England.

A Manual of Surgery. In Treatises by Various Authors. In three 12mo. volumes, containing 1866 pages, with 213 engravings. Price per volume, cloth, $2. See *Students' Series of Manuals,* page 4.

These volumes afford in a compact and portable form a complete view of the clinical aspects of modern surgery as understood and practised by thirty-three leading British surgeons.

BUTLIN, HENRY T., F. R. C. S.,
Assistant Surgeon to St. Bartholomew's Hospital, London.

Diseases of the Tongue. In one 12mo. volume of 456 pages, with 8 colored plates and 3 woodcuts. Cloth, $3.50. See *Series of Clinical Manuals,* page 4.

This book, the latest issue of the "Clinical Manuals for Practitioners and Students of Medicine," is a model of its kind. It is specially welcome, all the more so, since the text is really *illustrated* by a sufficient number of admirably executed colored plates. The work has been written by one whose opportunities have peculiarly fitted him for the task, since he teaches not only from a clinical but from a pathological standpoint. We heartily commend the book to our readers.—*The Medical News*, October 17, 1885.

ESMARCH, Dr. FRIEDRICH,
Professor of Surgery at the University of Kiel, etc.

Early Aid in Injuries and Accidents. Five Ambulance Lectures. Translated by H. R. H. PRINCESS CHRISTIAN. In one handsome small 12mo. volume of 109 pages, with 24 illustrations. Cloth, 75 cents.

TREVES, FREDERICK, F. R. C. S.,
Surgeon to and Lecturer on Surgery at the London Hospital.

Intestinal Obstruction. In one pocket-size 12mo. volume of 522 pages, with 60 illustrations. Limp cloth, blue edges, $2.00. See *Series of Clinical Manuals,* page 4.

A standard work on a subject that has not been so comprehensively treated by any contemporary English writer. Its completeness renders a full review difficult, since every chapter deserves minute attention, and it is impossible to do thorough justice to the author in a few paragraphs. *Intestinal Obstruction* is a work that will prove of equal value to the practitioner, the student, the pathologist, the physician and the operating surgeon.—*British Medical Journal*, Jan. 31, 1885.

BALL, CHARLES B., M. Ch., Dub., F. R. C. S. E.,
Surgeon and Teacher at Sir P. Dun's Hospital, Dublin.

Diseases of the Rectum and Anus. In one 12mo. volume of 550 pages. *Preparing.* See *Series of Clinical Manuals,* page 4.

HOLMES, TIMOTHY, M. A.,
Surgeon and Lecturer on Surgery at St. George's Hospital, London.

A System of Surgery; Theoretical and Practical. IN TREATISES BY VARIOUS AUTHORS. AMERICAN EDITION, THOROUGHLY REVISED AND RE-EDITED by JOHN H. PACKARD, M. D., Surgeon to the Episcopal and St. Joseph's Hospitals, Philadelphia, assisted by a corps of thirty-three of the most eminent American surgeons. In three large and very handsome imperial octavo volumes containing 3137 double-columned pages, with 979 illustrations on wood and 13 lithographic plates, beautifully colored. Price per volume, cloth, $6.00; leather, $7.00; half Russia, $7.50. Per set, cloth, $18.00; leather, $21.00; half Russia, $22.50. *Sold only by subscription.*

The authors of the original English edition are men of the front rank in England, and Dr. Packard has been fortunate in securing as his American coadjutors such men as Bartholow, Hyde, Hunt, Conner, Stimson, Morton, Hodgen, Jewell and their colleagues. As a whole, the work will be solid and substantial, and a valuable addition to the library of any medical man. It is more wieldly and more useful than the English edition, and with its companion work—"Reynolds' System of Medicine"—will well represent the present state of our science. One who is familiar with those two works will be fairly well furnished head-wise and hand-wise.—*The Medical News*, Jan. 7, 1882.

HAMILTON, FRANK H., M. D., LL. D.,
Surgeon to Bellevue Hospital, New York.

A Practical Treatise on Fractures and Dislocations. Seventh edition, thoroughly revised and much improved. In one very handsome octavo volume of 998 pages, with 379 illustrations. Cloth, $5.50; leather, $6.50; very handsome half Russia, open back, $7.00.

It is about twenty-five years ago since the first edition of this great work appeared. The edition now issued is the seventh, and this fact alone is enough to testify to the excellence of it in all particulars. Books upon special subjects do not usually command extended sale, but this one is without a rival in any language. It is essentially a practical treatise, and it gathers within its covers almost everything valuable that has been written about fractures and dislocations. The principles and methods of treatment are very fully given. The book is so well known that it does not require any lengthened review. We can only say that it is still unapproached as a treatise, and that it is a proof of the zeal and industry and great ability of its distinguished author.—*The Dublin Journal of Medical Science*, Feb. 1886.

With its first appearance in 1859, this work took rank among the classics in medical literature, and has ever since been quoted by surgeons the world over as an authority upon the topics of which it treats. The surgeon, if one can be found who does not already know the work, will find it scientific, forcible and scholarly in text, exhaustive in detail, and ever marked by a spirit of wise conservatism.—*Louisville Medical News*, Jan. 10, 1885.

STIMSON, LEWIS A., B. A., M. D.,
Professor of Pathological Anatomy at the University of the City of New York, Surgeon and Curator to Bellevue Hospital, Surgeon to the Presbyterian Hospital, New York, etc.

A Manual of Operative Surgery. New (second) edition. In one very handsome royal 12mo. volume of 503 pages, with 342 illustrations. Cloth, $2.50. *Just ready.*

Such works as this are sure to find large popularity when carefully prepared. This is certainly the case with the volume of Dr. Stimson. It is judiciously condensed, omitting nothing of much importance, and embracing a complete synopsis of the practical parts of surgery. The text will be found to represent in an entirely satisfactory manner the latest expressions of surgical science on its operative methods.—*Medical and Surgical Reporter*, Dec. 19, 1885.

By the same Author.

A Practical Treatise on Fractures. In one very handsome octavo volume of 598 pages, with 360 beautiful illustrations. Cloth, $4.75; leather, $5.75.

The author has given to the medical profession in this treatise on fractures what is likely to become a standard work on the subject. It is certainly not surpassed by any work written in the English, or, for that matter, any other language. The author tells us in a short, concise and comprehensive manner, all that is known about his subject. There is nothing scanty or superficial about it, as in most other treatises; on the contrary, everything is thorough. The chapters on repair of fractures and their treatment show him not only to be a profound student, but likewise a practical surgeon and pathologist. His mode of treatment of the different fractures is eminently sound and practical. We consider this work one of the best on fractures; and it will be welcomed not only as a text-book, but also by the surgeon in full practice.—*N. O. Medical and Surgical Journal*, March, 1883.

MARSH, HOWARD, F. R. C. S.,
Senior Assistant Surgeon to and Lecturer on Anatomy at St. Bartholomew's Hospital, London.

Diseases of the Joints. In one 12mo. volume. *Preparing.* See *Series of Clinical Manuals*, page 4.

PICK, T. PICKERING, F. R. C. S.,
Surgeon to and Lecturer on Surgery at St. George's Hospital, London.

Fractures and Dislocations. In one 12mo. volume of 530 pages, with 93 illustrations. Limp cloth, 2.00. *Just ready.* See *Series of Clinical Manuals*, page 4.

The author states that in writing the book he has kept the fact steadily in view that it should be essentially clinical, and he has therefore sought to present "a concise and practical treatise of the causes of the various common fractures and dislocations, the signs by which they may be recognized, and the appropriate treatment to be adopted for their cure." In this he has succeeded admirably. The book bears a distinctly clinical and practical stamp. In laying down rules as to symptoms and treatment, the author relies principally on those which he has found by practical experience to be most efficacious. The book contains an amount of information remarkable for one of its size.—*Boston Medical and Surgical Journal*, April 15, 1886.

BURNETT, CHARLES H., A. M., M. D.,
Professor of Otology in the Philadelphia Polyclinic; President of the American Otological Society.

The Ear, Its Anatomy, Physiology and Diseases. A Practical Treatise for the use of Medical Students and Practitioners. New (second) edition. In one handsome octavo volume of 580 pages, with 107 illustrations. Cloth, $4.00; leather, $5.00.

We note with pleasure the appearance of a second edition of this valuable work. When it first came out it was accepted by the profession as one of the standard works on modern aural surgery in the English language; and in his second edition Dr. Burnett has fully maintained his reputation, for the book is replete with valuable information and suggestions. The revision has been carefully carried out, and much new matter added. Dr. Burnett's work must be regarded as a very valuable contribution to aural surgery, not only on account of its comprehensiveness, but because it contains the results of the careful personal observation and experience of this eminent aural surgeon.—*London Lancet*, Feb. 21, 1885.

POLITZER, ADAM,
Imperial-Royal Prof. of Aural Therap. in the Univ. of Vienna.

A Text-Book of the Ear and its Diseases. Translated, at the Author's request, by JAMES PATTERSON CASSELLS, M. D., M. R. C. S. In one handsome octavo volume of 800 pages, with 257 original illustrations. Cloth, $5.50.

The work itself we do not hesitate to pronounce the best upon the subject of aural diseases which has ever appeared, systematic without being too diffuse on obsolete subjects, and eminently practical in every sense. The anatomical descriptions of each separate division of the ear are admirable, and profusely illustrated by woodcuts. They are followed immediately by the physiology of the section, and this again by the pathological physiology, an arrangement which serves to keep up the interest of the student by showing the direct application of what has preceded to the study of disease. The whole work can be recommended as a reliable guide to the student, and an efficient aid to the practitioner in his treatment.—*Boston Medical and Surgical Journal*, June 7, 1883.

JULER, HENRY E., F. R. C. S.,
Senior Ass't Surgeon, Royal Westminster Ophthalmic Hosp.; late Clinical Ass't, Moorfields, London.

A Handbook of Ophthalmic Science and Practice. In one handsome octavo volume of 460 pages, with 125 woodcuts, 27 colored plates, selections from the Test-types of Jaeger and Snellen, and Holmgren's Color-blindness Test. Cloth, $4.50; leather, $5.50.

This work is distinguished by the great number of colored plates which appear in it for illustrating various pathological conditions. They are very beautiful in appearance, and have been executed with great care as to accuracy. An examination of the work shows it to be one of high standing, one that will be regarded as an authority among ophthalmologists. The treatment recommended is such as the author has learned from actual experience to be the best.—*Cincinnati Medical News*, Dec. 1884.
It presents to the student concise descriptions and typical illustrations of all important eye affections, placed in juxtaposition, so as to be grasped at a glance. Beyond a doubt it is the best illustrated handbook of ophthalmic science which has ever appeared. Then, what is still better, these illustrations are nearly all original. We have examined this entire work with great care, and it represents the commonly accepted views of advanced ophthalmologists. We can most heartily commend this book to all medical students, practitioners and specialists. — *Detroit Lancet*, Jan. 1885.

NORRIS, WM. F., M. D., and OLIVER, CHAS. A., M. D.
Clin. Prof. of Ophthalmology in Univ. of Pa.

A Text-Book of Ophthalmology. In one octavo volume of about 500 pages, with illustrations. *Preparing.*

WELLS, J. SOELBERG, F. R. C. S.,
Professor of Ophthalmology in King's College Hospital, London, etc.

A Treatise on Diseases of the Eye. New (fifth) American from the third London edition. Thoroughly revised, with copious additions, by L. WEBSTER FOX, M. D. In one large octavo volume of about 850 pages, with about 275 illustrations on wood, six colored plates, and selections from the Test-types of Jaeger and Snellen. *Preparing.*

NETTLESHIP, EDWARD, F. R. C. S.,
Ophthalmic Surg. and Lect. on Ophth. Surg. at St. Thomas' Hospital, London.

The Student's Guide to Diseases of the Eye. Second edition. With a chapter on the Detection of Color-Blindness, by WILLIAM THOMSON, M. D., Ophthalmologist to the Jefferson Medical College. In one royal 12mo. volume of 416 pages, with 138 illustrations. Cloth, $2.00.

BROWNE, EDGAR A.,
Surgeon to the Liverpool Eye and Ear Infirmary and to the Dispensary for Skin Diseases.

How to Use the Ophthalmoscope. Being Elementary Instructions in Ophthalmoscopy, arranged for the use of Students. In one small royal 12mo. volume of 116 pages, with 35 illustrations. Cloth, $1.00.

LAWSON ON INJURIES TO THE EYE, ORBIT AND EYELIDS: Their Immediate and Remote Effects. 8 vo., 404 pp., 92 illus. Cloth, $3.50.
LAURENCE AND MOON'S HANDY BOOK OF OPHTHALMIC SURGERY, for the use of Practitioners. Second edition. In one octavo volume of 227 pages, with 65 illust. Cloth, $2.75.
CARTER'S PRACTICAL TREATISE ON DISEASES OF THE EYE. Edited by JOHN GREEN, M. D. In one handsome octavo volume.

ROBERTS, WILLIAM, M. D.,
Lecturer on Medicine in the Manchester School of Medicine, etc.

A Practical Treatise on Urinary and Renal Diseases, including Urinary Deposits. Fourth American from the fourth London edition. In one handsome octavo volume of 609 pages, with 81 illustrations. Cloth, $3.50.

The previous editions of this book have made it so familiar to and so highly esteemed by the medical public, that little more is necessary than a mere announcement of the appearance of this, their successor. But it is pleasant to be able to say that, good as those were, this is still better. In fact, we think it may be said to be the best book in print on the subject of which it treats.—*The American Journal of the Medical Sciences.*—Jan. 1886.

Among the numerous works on renal and urinary diseases now in circulation, perhaps Dr. Roberts' has the best claim to be regarded as "standard." The present edition shows evidence of having been carefully revised, and appears to be well up to the times. Dr. Roberts' book is an eminently useful and practical one, and we congratulate the author on its deserved popularity with the profession.—*Chicago Medical Journal and Examiner*, February, 1886.

The peculiar value and finish of the book are in a measure derived from its resolute maintenance of a clinical and practical character. It is an unrivalled exposition of everything which relates directly or indirectly to the diagnosis, prognosis and treatment of urinary diseases, and possesses a completeness not found elsewhere in our language in its account of the different affections.—*The Manchester Medical Chronicle*, July, 1885.

The work is practical in its character, and is regarded as an authority in the diseases which it treats. There is certainly no work that more fully sets forth the progress that has been made than this one of Dr. Roberts, and that more fully meets the wants of the physician in explaining the best methods of treatment. We have no hesitation in recommending it to our subscribers.—*Cincinnati Medical News*, June, 1885.

PURDY, CHARLES W., A. M., M. D.
Bright's Disease and Allied Disorders. In one octavo volume of 350 pages, with illustrations. *Shortly.*

MORRIS, HENRY, M. B., F. R. C. S.,
Surgeon to and Lecturer on Surgery at Middlesex Hospital, London.

Surgical Diseases of the Kidney. In one 12mo. volume of 554 pages, with 40 woodcuts, and 6 colored plates. *Just ready.* Limp cloth, $2.25. See *Series of Clinical Manuals*, page 4.

We highly approve of Mr. Morris's book and strongly recommend it to practical surgeons.—*Edinburgh Medical Journal*, April, 1886.

LUCAS, CLEMENT, M. B., B. S., F. R. C. S.,
Senior Assistant Surgeon to Guy's Hospital, London.

Diseases of the Urethra. In one 12mo. volume. *Preparing.* See *Series of Clinical Manuals*, page 4.

THOMPSON, SIR HENRY,
Surgeon and Professor of Clinical Surgery to University College Hospital, London.

Lectures on Diseases of the Urinary Organs. Second American from the third English edition. In one 8vo. volume of 203 pp., with 25 illustrations. Cloth, $2.25.

By the Same Author.
On the Pathology and Treatment of Stricture of the Urethra and Urinary Fistulæ. From the third English edition. In one octavo volume of 359 pages, with 47 cuts and 3 plates. Cloth, $3.50.

AN AMERICAN SYSTEM OF DENTISTRY.
A System of Dentistry, in Treatises by Various Authors. Edited by WILBUR F. LITCH, M. D., D. D. S., Professor of Prosthetic Dentistry, Materia Medica and Therapeutics in the Pennsylvania College of Dental Surgery. In three very handsome octavo volumes of about 800 pages each, richly illustrated. Per volume, cloth, $6; leather, $7; half Morocco, gilt top, $8. Volume I., *just ready. For sale by subscription only.*

COLEMAN, A., L. R. C. P., F. R. C. S., Exam. L. D. S.,
Senior Dent. Surg. and Lect. on Dent. Surg. at St. Bartholomew's Hosp. and the Dent. Hosp., London.

A Manual of Dental Surgery and Pathology. Thoroughly revised and adapted to the use of American Students, by THOMAS C. STELLWAGEN, M. A., M. D., D. D. S., Prof. of Physiology at the Philadelphia Dental College. In one handsome octavo volume of 412 pages, with 331 illustrations. Cloth, $3.25.

This volume presents a highly creditable appearance, and deserves to rank among the most important of recent contributions to dental literature. Mr. Coleman has presented his methods of practice, for the most part, in a plain and concise manner, and the work of the American editor has been conscientiously performed. He has evidently labored to present his convictions of the best modes of practice for the instruction of those commencing a professional career, and he has faithfully endeavored to teach to others all that he has acquired by his own observation and experience. The book deserves a place in the library of every dentist.—*Dental Cosmos*, May, 1882.

It should be in the possession of every practitioner in this country. The part devoted to first and second dentition and irregularities in the permanent teeth is fully worth the price. In fact, price should not be considered in purchasing such a work. If the money you spend of our so-called standard text-books could be converted into such publications as this, much good would result.—*Southern Dental Journal*, May, 1882.

BASHAM ON RENAL DISEASES: A Clinical Guide to their Diagnosis and Treatment. In one 12mo. vol. of 304 pages, with 21 illustrations. Cloth, $2.00.

BUMSTEAD, F. J., M. D., LL. D., and TAYLOR, R. W., A. M., M. D.,
Late Professor of Venereal Diseases at the College of Physicians and Surgeons, New York, etc. Surgeon to Charity Hospital, New York, Prof. of Venereal and Skin Diseases in the University of Vermont, Pres. of the Am. Dermatological Ass'n.

The Pathology and Treatment of Venereal Diseases. Including the results of recent investigations upon the subject. Fifth edition, revised and largely rewritten, by Dr. Taylor. In one large and handsome octavo volume of 898 pages with 139 illustrations, and thirteen chromo-lithographic figures. Cloth, $4.75; leather, $5.75; very handsome half Russia, $6.25.

It is a splendid record of honest labor, wide research, just comparison, careful scrutiny and original experience, which will always be held as a high credit to American medical literature. This is not only the best work in the English language upon the subjects of which it treats, but also one which has no equal. In other tongues for its clear, comprehensive and practical handling of its themes.—*American Journal of the Medical Sciences*, Jan., 1884.

It is certainly the best single treatise on venereal in our own, and probably the best in any language.—*Boston Medical and Surgical Journal*, April 3, 1884.

The character of this standard work is so well known that it would be superfluous here to pass in review its general or special points of excellence. The verdict of the profession has been passed; it has been accepted as the most thorough and complete exposition of the pathology and treatment of venereal diseases in the language. Admirable as a model of clear description, an exponent of sound pathological doctrine, and a guide for rational and successful treatment, it is an ornament to the medical literature of this country. The additions made to the present edition are eminently judicious, from the standpoint of practical utility.—*Journal of Cutaneous and Venereal Diseases*, Jan. 1884.

CORNIL, V.,
Professor to the Faculty of Medicine of Paris, and Physician to the Lourcine Hospital.

Syphilis, its Morbid Anatomy, Diagnosis and Treatment. Specially revised by the Author, and translated with notes and additions by J. HENRY C. SIMES, M. D., Demonstrator of Pathological Histology in the University of Pennsylvania, and J. WILLIAM WHITE, M. D., Lecturer on Venereal Diseases and Demonstrator of Surgery in the University of Pennsylvania. In one handsome octavo volume of 461 pages, with 84 very beautiful illustrations. Cloth, $3.75.

The anatomical and histological characters of the hard and soft sore are admirably described. The multiform cutaneous manifestations of the disease are dealt with histologically in a masterly way, as we should indeed expect them to be, and the accompanying illustrations are executed carefully and well. The various nervous lesions which are the recognized outcome of the syphilitic dyscrasia are treated with care and consideration. Syphilitic epilepsy, paralysis, cerebral syphilis and locomotor ataxia are subjects full of interest; and nowhere in the whole volume is the clinical experience of the author or the wide acquaintance of the translators with medical literature more evident. The anatomy, the histology, the pathology and the clinical features of syphilis are represented in this work in their best, most practical and most instructive form, and no one will rise from its perusal without the feeling that his grasp of the wide and important subject on which it treats is a stronger and surer one.—*The London Practitioner*, Jan. 1882.

HUTCHINSON, JONATHAN, F. R. S., F. R. C. S.,
Consulting Surgeon to the London Hospital.

Syphilis. In one 12mo. volume. *Preparing*. See Series of Clinical Manuals, page 4.

GROSS, SAMUEL W., A. M., M. D.,
Professor of the Principles of Surgery and of Clinical Surgery in the Jefferson Medical College of Phila.

A Practical Treatise on Impotence, Sterility, and Allied Disorders of the Male Sexual Organs. Second edition, thoroughly revised. In one very handsome octavo volume of 168 pages, with 16 illustrations. Cloth, $1.50.

The author of this monograph is a man of positive convictions and vigorous style. This is justified by his experience and by his study, which has gone hand in hand with his experience. In regard to the various organic and functional disorders of the male generative apparatus, he has had exceptional opportunities for observation, and his book shows that he has not neglected to compare his own views with those of other authors. The result is a work which can be safely recommended to both physicians and surgeons as a guide in the treatment of the disturbances it refers to. It is the best treatise on the subject with which we are acquainted.—*The Medical News*, Sept. 1, 1883.

GROSS, S. D., M. D., LL. D., D. C. L., etc.

A Practical Treatise on the Diseases, Injuries and Malformations of the Urinary Bladder, the Prostate Gland and the Urethra. Third edition, thoroughly revised by SAMUEL W. GROSS, M. D. In one octavo volume of 574 pages, with 170 illustrations. Cloth, $4.50.

CULLERIER, A., & BUMSTEAD, F. J., M.D., LL.D.,
Surgeon to the Hôpital du Midi. Late Professor of Venereal Diseases in the College of Physicians and Surgeons, New York.

An Atlas of Venereal Diseases. Translated and edited by FREEMAN J. BUMSTEAD, M. D. In one imperial 4to. volume of 328 pages, double-columns, with 26 plates, containing about 150 figures, beautifully colored, many of them the size of life. Strongly bound in cloth, $17.00. A specimen of the plates and text sent by mail, on receipt of 25 cts.

HILL ON SYPHILIS AND LOCAL CONTAGIOUS DISORDERS. In one 8vo vol. of 479 p. Cloth, $3.25.
LEE'S LECTURES ON SYPHILIS AND SOME FORMS OF LOCAL DISEASE AFFECTING PRINCIPALLY THE ORGANS OF GENERATION. In one 8vo. vol. of 246 pages. Cloth, $2.25.

HYDE, J. NEVINS, A. M., M. D.,
Professor of Dermatology and Venereal Diseases in Rush Medical College, Chicago.

A Practical Treatise on Diseases of the Skin. For the use of Students and Practitioners. In one handsome octavo volume of 570 pages, with 66 beautiful and elaborate illustrations. Cloth, $4.25; leather, $5.25.

The author has given the student and practitioner a work admirably adapted to the wants of each. We can heartily commend the book as a valuable addition to our literature and a reliable guide to students and practitioners in their studies and practice.—*Am. Journ. of Med. Sci.*, July, 1883.

Especially to be praised are the practical suggestions as to what may be called the common-sense treatment of eczema. It is quite impossible to exaggerate the judiciousness with which the formulæ for the external treatment of eczema are selected, and what is of equal importance, the full and clear instructions for their use.—*London Medical Times and Gazette*, July 28, 1883.

The work of Dr. Hyde will be awarded a high position. The student of medicine will find it peculiarly adapted to his wants. Notwithstanding the extent of the subject to which it is devoted, yet it is limited to a single and not very large volume, without omitting a proper discussion of the topics. The conciseness of the volume, and the setting forth of only what can be held as facts will also make it acceptable to general practitioners.—*Cincinnati Medical News*, Feb. 1883.

The aim of the author has been to present to his readers a work not only expounding the most modern conceptions of his subject, but presenting what is of standard value. He has more especially devoted its pages to the treatment of disease, and by his detailed descriptions of therapeutic measures has adapted them to the needs of the physician in active practice. In dealing with these questions the author leaves nothing to the presumed knowledge of the reader, but enters thoroughly into the most minute description, so that one is not only told what should be done under given conditions but how to do it as well. It is therefore in the best sense "a practical treatise." That it is comprehensive, a glance at the index will show.—*Maryland Medical Journal*, July 7, 1883.

Professor Hyde has long been known as one of the most intelligent and enthusiastic representatives of dermatology in the west. His numerous contributions to the literature of this specialty have gained for him a favorable recognition as a careful, conscientious and original observer. The remarkable advances made in our knowledge of diseases of the skin, especially from the standpoint of pathological histology and improved methods of treatment, necessitate a revision of the older text-books at short intervals in order to bring them up to the standard demanded by the march of science. This last contribution of Dr. Hyde is an effort in this direction. He has attempted, as he informs us, the task of presenting in a condensed form the results of the latest observation and experience. A careful examination of the work convinces us that he has accomplished his task with painstaking fidelity and with a creditable result.—*Journal of Cutaneous and Venereal Diseases*, June, 1883.

FOX, T., M.D., F.R.C.P., and FOX, T.C., B.A., M.R.C.S.,
Physician to the Department for Skin Diseases, University College Hospital, London.   *Physician for Diseases of the Skin to the Westminster Hospital, London.*

An Epitome of Skin Diseases. With Formulæ. For Students and Practitioners. Third edition, revised and enlarged. In one very handsome 12mo. volume of 238 pages. Cloth, $1.25.

The third edition of this convenient handbook calls for notice owing to the revision and expansion which it has undergone. The arrangement of skin diseases in alphabetical order, which is the method of classification adopted in this work, becomes a positive advantage to the student. The book is one which we can strongly recommend, not only to students but also to practitioners who require a compendious summary of the present state of dermatology.—*British Medical Journal*, July 2, 1883.

We cordially recommend Fox's *Epitome* to those whose book is limited and who wish a handy manual to lie upon the table for instant reference. Its alphabetical arrangement is suited to this use, for all one has to know is the name of the disease, and here are its description and the appropriate treatment at hand and ready for instant application. The present edition has been very carefully revised and a number of new diseases are described, while most of the recent additions to dermal therapeutics find mention, and the formulary at the end of the book has been considerably augmented.—*The Medical News*, December, 1883.

MORRIS, MALCOLM, F. R. C. S.,
Joint Lecturer on Dermatology at St. Mary's Hospital Medical School, London.

Skin Diseases; Including their Definitions, Symptoms, Diagnosis, Prognosis, Morbid Anatomy and Treatment. A Manual for Students and Practitioners. In one 12mo. volume of 316 pages, with illustrations. Cloth, $1.75.

To physicians who would like to know something about skin diseases, so that when a patient presents himself for relief they can make a correct diagnosis and prescribe a rational treatment, we unhesitatingly recommend this little book of Dr. Morris. The affections of the skin are described in a terse, lucid manner, and their several characteristics so plainly set forth that diagnosis will be easy. The treatment in each case is such as the experience of the most eminent dermatologists advises.—*Cincinnati Medical News*, April, 1880.

This is emphatically a learner's book; for we can safely say, that in the whole range of medical literature there is no book of a like scope which for clearness of expression and methodical arrangement is better adapted to promote a rational conception of dermatology—a branch confessedly difficult and perplexing to the beginner.—*St. Louis Courier of Medicine*, April, 1880.

The writer has certainly given in a small compass a large amount of well-compiled information, and his little book compares favorably with any other which has emanated from England, while in many points he has emancipated himself from the stubbornly adhered to errors of others of his countrymen. There is certainly excellent material in the book which will well repay perusal.—*Boston Med. and Surg. Journ.*, March, 1880.

WILSON, ERASMUS, F. R. S.
The Student's Book of Cutaneous Medicine and Diseases of the Skin. In one handsome small octavo volume of 535 pages. Cloth, $3.50.

HILLIER, THOMAS, M. D.,
Physician to the Skin Department of University College, London.

Handbook of Skin Diseases; for Students and Practitioners. Second American edition. In one 12mo. volume of 353 pages, with plates. Cloth, $2.25.

AN AMERICAN SYSTEM OF GYNÆCOLOGY.

A System of Gynæcology, in Treatises by Various Authors. Edited by MATTHEW D. MANN, M. D., Professor of Obstetrics and Gynæcology in the University of Buffalo, N. Y. In two handsome octavo volumes, richly illustrated. *In active preparation.*

LIST OF CONTRIBUTORS.

WILLIAM H. BAKER, M. D.,
FORDYCE BARKER, M. D.,
ROBERT BATTEY, M. D.,
SAMUEL C. BUSEY, M. D.,
HENRY F. CAMPBELL, M. D.,
HENRY C. COE, M. D.,
E. C. DUDLEY, M. D.,
GEORGE J. ENGELMANN, M. D.,
HENRY F. GARRIGUES, M. D.,
WILLIAM GOODELL, M. D.,
EGBERT H. GRANDIN, M. D.,
SAMUEL W. GROSS, M. D.,
JAMES B. HUNTER, M. D.,
A. REEVES JACKSON, M. D.,

EDWARD W. JENKS, M. D.,
WILLIAM T. LUSK, M. D.,
MATTHEW D. MANN, M. D.,
ROBERT B. MAURY, M. D.,
PAUL F. MUNDÉ, M. D.,
C. D. PALMER, M. D.,
WILLIAM M. POLK, M. D.,
THADDEUS A. REAMY, M. D.,
A. D. ROCKWELL, M. D.,
ALEX. J. C. SKENE, M. D.,
R. STANSBURY SUTTON, A. M., M. D.,
T. GAILLARD THOMAS, M. D.,
ELY VAN DE WARKER, M. D.,
W. GILL WYLIE, M. D.

THOMAS, T. GAILLARD, M. D.,
Professor of Diseases of Women in the College of Physicians and Surgeons, N. Y.

A Practical Treatise on the Diseases of Women. Fifth edition, thoroughly revised and rewritten. In one large and handsome octavo volume of 810 pages, with 266 illustrations. Cloth, $5.00; leather, $6.00; very handsome half Russia, raised bands, $6.50.

The words which follow "fifth edition" are in this case no mere formal announcement. The alterations and additions which have been made are both numerous and important. The attraction and the permanent character of this book lie in the clearness and truth of the clinical descriptions of diseases; the fertility of the author in therapeutic resources and the fulness with which the details of treatment are described; the definite character of the teaching; and last, but not least, the evident candor which pervades it. We would also particularize the fulness with which the history of the subject is gone into, which makes the book additionally interesting and gives it value as a work of reference.—*London Medical Times and Gazette*, July 30, 1881.

The determination of the author to keep his book foremost in the rank of works on gynæcology is most gratifying. Recognizing the fact that this can only be accomplished by frequent and thorough revision, he has spared no pains to make the present edition more desirable even than the previous one. As a book of reference for the busy practitioner it is unequalled.—*Boston Medical and Surgical Journal*, April 7, 1880.

It has been enlarged and carefully revised. It is a condensed encyclopædia of gynæcological medicine. The style of arrangement, the masterly manner in which each subject is treated, and the honest convictions derived from probably the largest clinical experience in that specialty of any in this country, all serve to commend it in the highest terms to the practitioner.—*Nashville Jour. of Med. and Surg.*, Jan. 1881.

That the previous editions of the treatise of Dr. Thomas were thought worthy of translation into German, French, Italian and Spanish, is enough to give it the stamp of genuine merit. At home it has made its way into the library of every obstetrician and gynæcologist as a safe guide to practice. No small number of additions have been made to the present edition to make it correspond to recent improvements in treatment.—*Pacific Medical and Surgical Journal*, Jan. 1881.

EDIS, ARTHUR W., M. D., Lond., F. R. C. P., M. R. C. S.,
Assist. Obstetric Physician to Middlesex Hospital, late Physician to British Lying-in Hospital.

The Diseases of Women. Including their Pathology, Causation, Symptoms, Diagnosis and Treatment. A Manual for Students and Practitioners. In one handsome octavo volume of 576 pages, with 148 illustrations. Cloth, $3.00; leather, $4.00.

It is a pleasure to read a book so thoroughly good as this one. The special qualities which are conspicuous are thoroughness in covering the whole ground, clearness of description and conciseness of statement. Another marked feature of the book is the attention paid to the details of many minor surgical operations and procedures, as, for instance, the use of tents, application of leeches, and use of hot water injections. These are among the more common methods of treatment, and yet very little is said about them in many of the text-books. The book is one to be warmly recommended especially to students and general practitioners, who need a concise but complete *résumé* of the whole subject. Specialists, too, will find many useful hints in its pages.—*Boston Med. and Surg. Journ.*, March 2, 1882.

The greatest pains have been taken with the sections relating to treatment. A liberal selection of remedies is given for each morbid condition, the strength, mode of application and other details being fully explained. The descriptions of gynæcological manipulations and operations are full, clear and practical. Much care has also been bestowed on the parts of the book which deal with diagnosis—we note especially the pages dealing with the differentiation, one from another, of different kinds of abdominal tumors. The practitioner will therefore find in this book the kind of knowledge he most needs in his daily work, and he will be pleased with the clearness and fulness of the information there given.—*The Practitioner*, Feb. 1882.

BARNES, ROBERT, M. D., F. R. C. P.,
Obstetric Physician to St. Thomas' Hospital, London, etc.

A Clinical Exposition of the Medical and Surgical Diseases of Women. In one handsome octavo volume, with numerous illustrations. New edition. *Preparing.*

WEST, CHARLES, M. D.

Lectures on the Diseases of Women. Third American from the third London edition. In one octavo volume of 543 pages. Cloth, $3.75; leather, $4.75.

EMMET, THOMAS ADDIS, M. D., LL. D.,
Surgeon to the Woman's Hospital, New York, etc.

The Principles and Practice of Gynæcology; For the use of Students and Practitioners of Medicine. New (third) edition, thoroughly revised. In one large and very handsome octavo volume of 880 pages, with 150 illustrations. Cloth, $5; leather, $6; very handsome half Russia, raised bands, $6.50.

We are in doubt whether to congratulate the author more than the profession upon the appearance of the third edition of this well-known work. Embodying, as it does, the life-long experience of one who has conspicuously distinguished himself as a bold and successful operator, and who has devoted so much attention to the specialty, we feel sure the profession will not fail to appreciate the privilege thus offered them of perusing the views and practice of the author. His earnestness of purpose and conscientiousness are manifest. He gives not only his individual experience but endeavors to represent the actual state of gynæcological science and art.—*British Medical Journal*, May 10, 1885.

No jot or tittle of the high praise bestowed upon the first edition is abated. It is still a book of marked personality, one based upon large clinical experience, containing large and valuable additions to our knowledge, evidently written not only with honesty of purpose, but with a conscientious sense of responsibility, and a book that is at once a credit to its author and to American medical literature. We repeat that it is a book to be studied, and one that is indispensable to every practitioner giving any attention to gynæcology.—*American Journal of the Medical Sciences*, April, 1885.

The time has passed when Emmet's *Gynæcology* was to be regarded as a book for a single country or for a single generation. It has always been his aim to popularize gynæcology, to bring it within easy reach of the general practitioner. The originality of the ideas, aside from the perfect confidence which we feel in the author's statements, compels our admiration and respect. We may well take an honest pride in Dr. Emmet's work and feel that his book can hold its own against the criticism of two continents. It represents all that is most earnest and most thoughtful in American gynæcology. Emmet's work will continue to reflect the individuality, the sterling integrity and the kindly heart of its honored author long after smaller books have been forgotten.—*American Journal of Obstetrics*, May, 1885.

DUNCAN, J. MATTHEWS, M.D., LL. D., F. R. S. E., etc.

Clinical Lectures on the Diseases of Women; Delivered in Saint Bartholomew's Hospital. In one handsome octavo volume of 175 pages. Cloth, $1.50.

They are in every way worthy of their author; indeed, we look upon them as among the most valuable of his contributions. They are all upon matters of great interest to the general practitioner. Some of them deal with subjects that are not, as a rule, adequately handled in the text-books; others of them, while bearing upon topics that are usually treated of at length in such works, yet bear such a stamp of individuality that they deserve to be widely read.—*N. Y. Medical Journal*, March, 1880.

MAY, CHARLES H., M. D.
Late House Surgeon to Mount Sinai Hospital, New York.

A Manual of the Diseases of Women. Being a concise and systematic exposition of the theory and practice of gynæcology. In one 12mo. volume of 342 pages. Cloth, $1.75. *Just ready.*

Medical students will find this work adapted to their wants. Also practitioners of medicine will find it exceedingly convenient to consult for the purpose of refreshing their minds upon the leading points of a gynæcological subject. By systematic condensation, the omission of disputed questions, and the presentation only of accepted views, it constitutes a very satisfactory exposition of the leading principles of gynæcology as they are understood at the present time.—*Cincinnati Medical News*, Nov. 1885.

HODGE, HUGH L., M. D.,
Emeritus Professor of Obstetrics, etc., in the University of Pennsylvania.

On Diseases Peculiar to Women; Including Displacements of the Uterus. Second edition, revised and enlarged. In one beautifully printed octavo volume of 519 pages, with original illustrations. Cloth, $4.50.

By the Same Author.

The Principles and Practice of Obstetrics. Illustrated with large lithographic plates containing 159 figures from original photographs, and with numerous woodcuts. In one large quarto volume of 542 double-columned pages. Strongly bound in cloth, $14.00. Specimens of the plates and letter-press will be forwarded to any address, free by mail, on receipt of six cents in postage stamps.

RAMSBOTHAM, FRANCIS H., M. D.

The Principles and Practice of Obstetric Medicine and Surgery; In reference to the Process of Parturition. A new and enlarged edition, thoroughly revised by the Author. With additions by W. V. KEATING, M. D., Professor of Obstetrics, etc., in the Jefferson Medical College of Philadelphia. In one large and handsome imperial octavo volume of 640 pages, with 64 full-page plates and 43 woodcuts in the text, containing in all nearly 200 beautiful figures. Strongly bound in leather, with raised bands, $7.

WINCKEL, F.

A Complete Treatise on the Pathology and Treatment of Childbed, For Students and Practitioners. Translated, with the consent of the Author, from the second German edition, by J. R. CHADWICK, M. D. Octavo 484 pages. Cloth, $4.00.

ASHWELL'S PRACTICAL TREATISE ON THE DISEASES PECULIAR TO WOMEN. Third American from the third and revised London edition. In one 8vo. vol., pp. 520. Cloth, $3.50.
CHURCHILL ON THE PUERPERAL FEVER AND OTHER DISEASES PECULIAR TO WOMEN. In one 8vo. vol. of 464 pages. Cloth, $2.50.
MEIGS ON THE NATURE, SIGNS AND TREATMENT OF CHILDBED FEVER. In one 8vo. volume of 346 pages. Cloth, $2.00.

BARNES, ROBERT, M. D., and FANCOURT, M. D.,
Phys. to the General Lying-in Hosp., Lond. *Obstetric Phys. to St. Thomas' Hosp., Lond.*

A System of Obstetric Medicine and Surgery, Theoretical and Clinical. For the Student and the Practitioner. The Section on Embryology contributed by Prof. Milnes Marshall. In one handsome octavo volume of 872 pages, with 231 illustrations. Cloth, $5; leather, $6. *Just ready.*

This system will be eagerly sought for, not only on account of its intrinsic merit, but also because the reputation which the elder Barnes, in particular, has secured, carries with it the conviction that any book emanating from him is necessarily sound in teaching and conservative in practice. It is indeed eminently fitting that a man who has done so much towards systematizing the obstetric art, who for so many years has been widely known as a capable teacher and trusted accoucheur, should embody within a single treatise the system which he has taught and in practice tested, and which is the outcome of a lifetime of earnest labor, careful observation and deep study. The result of this arrangement is the production of a work which rises above criticism and which in no respect need yield the palm to any obstetrical treatise hitherto published.—*American Journal of Obstetrics,* Feb. 1886.

PLAYFAIR, W. S., M. D., F. R. C. P.,
Professor of Obstetric Medicine in King's College, London, etc.

A Treatise on the Science and Practice of Midwifery. New (fourth) American, from the fifth English edition. Edited, with additions, by ROBERT P. HARRIS, M. D. In one handsome octavo volume of 654 pages, with 3 plates and 201 engravings Cloth, $4; leather, $5; half Russia, $5.50. *Just ready.*

This still remains a favorite in America, not only because the author is recognized as a safe guide and eminently progressive man, but also as sparing no effort to make each successive edition a faithful mirror of the latest and best practice. A work so frequently noticed as the present requires no further review. We believe that this edition is simply the forerunner of many others, and that the demand will keep pace with the supply.—*American Journal of Obstetrics,* Nov. 1885.

Since its first publication, only eight years ago, it has rapidly become the favorite text-book, to the practical exclusion of all others. A large measure of its popularity is due to the clear and easy style in which it is written. Few text-books for students have very much to boast of in this respect.—*Medical Record.*

In the short time that this excellent and highly esteemed work has been before the profession it has reached a fourth edition in this country and a fifth one in England. This fact alone speaks in high praise of it, and it seems to us that scarcely more need be said of it in the way of endorsement of its value. As a text book for students and for work on obstetrics superior to the work of Dr. Playfair. Its teachings are practical, written in plain language, and afford a better understanding of the art of midwifery. No one can be disappointed in it.—*Cincinnati Medical News,* June, 1886.

BARKER, FORDYCE, A. M., M. D., LL. D. Edin.,
Clinical Professor of Midwifery and the Diseases of Women in the Bellevue Hospital Medical College, New York, Honorary Fellow of the Obstetrical Societies of London and Edinburgh, etc., etc.

Obstetrical and Clinical Essays. In one handsome 12mo. volume of about 300 pages. *Preparing.*

KING, A. F. A., M. D.,
Professor of Obstetrics and Diseases of Women in the Medical Department of the Columbian University, Washington, D. C., and in the University of Vermont, etc.

A Manual of Obstetrics. Second edition. In one very handsome 12mo. volume of 331 pages, with 59 illustrations. Cloth, $2.00.

It must be acknowledged that this is just what it pretends to be—a sound guide, a portable epitome, a work in which only indispensable matter has been presented, leaving out all padding and chaff, and one in which the student will find pure wheat or condensed nutriment.—*New Orleans Medical and Surgical Journal,* May, 1884.

In a series of short paragraphs and by a condensed style of composition, the writer has presented a great deal of what it is well that every obstetrician should know and be ready to practice or prescribe. The fact that the demand for the volume has been such as to exhaust the first edition in a little over a year and a half speaks well for its popularity.—*American Journal of the Medical Sciences,* April, 1884.

BARNES, FANCOURT, M. D.,
Obstetric Physician to St. Thomas' Hospital, London.

A Manual of Midwifery for Midwives and Medical Students. In one royal 12mo. volume of 197 pages, with 50 illustrations. Cloth, $1.25.

PARVIN, THEOPHILUS, M. D., LL. D.,
Professor of Obstetrics and the Diseases of Women and Children in the Jefferson Medical College.

A Treatise on Midwifery. In one very handsome octavo volume of about 550 pages, with numerous illustrations. *Preparing.*

PARRY, JOHN S., M. D.,
Obstetrician to the Philadelphia Hospital, Vice-President of the Obstet. Society of Philadelphia.

Extra-Uterine Pregnancy: Its Clinical History, Diagnosis, Prognosis and Treatment. In one handsome octavo volume of 272 pages. Cloth, $2.50.

TANNER, THOMAS HAWKES, M. D.
On the Signs and Diseases of Pregnancy. First American from the second English edition. Octavo, 490 pages, with 4 colored plates and 16 woodcuts. Cloth, $4.25.

LEISHMAN, WILLIAM, M. D.,
Regius Professor of Midwifery in the University of Glasgow, etc.

A System of Midwifery, Including the Diseases of Pregnancy and the Puerperal State. Third American edition, revised by the Author, with additions by JOHN S. PARRY, M. D., Obstetrician to the Philadelphia Hospital, etc. In one large and very handsome octavo volume of 740 pages, with 205 illustrations. Cloth, $4.50; leather, $5.50; very handsome half Russia, raised bands, $6.00.

The author is broad in his teachings, and discusses briefly the comparative anatomy of the pelvis and the mobility of the pelvic articulations. The second chapter is devoted especially to the study of the pelvis, while in the third the female organs of generation are introduced. The structure and development of the ovum are admirably described. Then follow chapters upon the various subjects embraced in the study of midwifery. The descriptions throughout the work are plain and pleasing. It is sufficient to state that in this, the last edition of this well-known work, every recent advancement in this field has been brought forward.—*Physician and Surgeon*, Jan. 1880.

To the American student the work before us must prove admirably adapted. Complete in all its parts, essentially modern in its teachings, and with demonstrations noted for clearness and precision, it will gain in favor and be recognized as a work of standard merit. The work cannot fail to be popular and is cordially recommended.—*N. O. Med. and Surg. Journ.*, March, 1880.

LANDIS, HENRY G., A. M., M. D.,
Professor of Obstetrics and the Diseases of Women in Starling Medical College, Columbus, O.

The Management of Labor, and of the Lying-in Period. In one handsome 12mo. volume of 334 pages, with 28 illustrations. Cloth, $1.75. *Just ready.*

This is a book we can heartily recommend. The author goes much more practically into the details of the management of labor than most text-books, and is so readable throughout as to tempt any one who should happen to commence the book to read it through. The author presupposes a theoretical knowledge of obstetrics, and has consistently excluded from this little work everything that is not of practical use in the lying-in room. We think that if it is as widely read as it deserves, it will do much to improve obstetric practice in general.—*New Orleans Medical and Surgical Journal*, Mar. 1880.

SMITH, J. LEWIS, M. D.,
Clinical Professor of Diseases of Children in the Bellevue Hospital Medical College, N. Y.

A Treatise on the Diseases of Infancy and Childhood. New (sixth) edition, thoroughly revised and rewritten. In one handsome octavo volume of 867 pages, with 40 illustrations. Cloth, $4.50; leather, $5.50; half Russia, $6.00. *Just ready.*

No better work on children's diseases could be placed in the hands of the student, containing, as it does, a very complete account of the symptoms and pathology of the diseases of early life, and possessing the further advantage, in which it stands alone amongst other works on its subject, of recommending treatment in accordance with the most recent therapeutical views.—*British and Foreign Medico-Chirurgical Review.*

Among American medical text-books which have become classic, Smith on Children may be ranked with the foremost. The author has continuously kept in view the eminently practical character of his work, which made it so popular in former editions. A very commendable feature is the increasing space devoted to therapeutics, in which the author, besides drawing on his own rich mine of clinical experience, gives in addition, the most improved forms of treatment as gleaned from the works of others.—*Cincinnati Lancet and Clinic*, Feb. 27, 1886.

KEATING, JOHN M., M. D.,
Lecturer on the Diseases of Children at the University of Pennsylvania, etc.

The Mother's Guide in the Management and Feeding of Infants. In one handsome 12mo. volume of 118 pages. Cloth, $1.00.

Works like this one will aid the physician immensely, for it saves the time he is constantly giving his patients in instructing them on the subjects here dwelt upon so thoroughly and practically. Dr. Keating has written a practical book, has carefully avoided unnecessary repetition, and successfully instructed the mother in such details of the treatment of her child as devolve upon her. He has studiously omitted giving prescriptions, and instructs the mother when to call upon the doctor, as his duties are totally distinct from hers. —*American Journal of Obstetrics*, October, 1881.

OWEN, EDMUND, M. B., F. R. C. S.,
Surgeon to the Children's Hospital, Great Ormond St., London.

Surgical Diseases of Children. In one 12mo. volume of 525 pages, with 4 chromo-lithographic plates and 85 woodcuts. *Just ready.* Cloth, $2. See *Series of Clinical Manuals*, page 4.

We look with considerable interest at the work, coming as it does from the hands of a surgeon of special experience in this subject and recognized as an able teacher as well as a peculiarly practical surgeon. It certainly may be looked to as the type of a practical manual.—*London Medical Record*, January 15, 1886.

WEST, CHARLES, M. D.,
Physician to the Hospital for Sick Children, London, etc.

Lectures on the Diseases of Infancy and Childhood. Fifth American from 6th English edition. In one octavo volume of 686 pages. Cloth, $4.50; leather, $5.50.

By the Same Author.

On Some Disorders of the Nervous System in Childhood. In one small 12mo. volume of 127 pages. Cloth, $1.00.

CONDIE'S PRACTICAL TREATISE ON THE DISEASES OF CHILDREN. Sixth edition, revised and augmented. In one octavo volume of 779 pages. Cloth, $5.25; leather, $6.25.

TIDY, CHARLES MEYMOTT, M. B., F. C. S.,
Professor of Chemistry and of Forensic Medicine and Public Health at the London Hospital, etc.

Legal Medicine. VOLUME II. Legitimacy and Paternity, Pregnancy, Abortion, Rape, Indecent Exposure, Sodomy, Bestiality, Live Birth, Infanticide, Asphyxia, Drowning, Hanging, Strangulation, Suffocation. Making a very handsome imperial octavo volume of 529 pages. Cloth, $6.00; leather, $7.00.

VOLUME I. Containing 664 imperial octavo pages, with two beautiful colored plates. Cloth, $6.00; leather, $7.00.

The satisfaction expressed with the first portion of this work is in no wise lessened by a perusal of the second volume. We find it characterized by the same fulness of detail and clearness of expression which we had occasion so highly to commend in our former notice, and which render it so valuable to the medical jurist. The copious tables of cases appended to each division of the subject, must have cost the author a prodigious amount of labor and research, but they constitute one of the most valuable features of the book, especially for reference in medico-legal trials.—*American Journal of the Medical Sciences,* April, 1884.

TAYLOR, ALFRED S., M. D.,
Lecturer on Medical Jurisprudence and Chemistry in Guy's Hospital, London.

A Manual of Medical Jurisprudence. Eighth American from the tenth London edition, thoroughly revised and rewritten. Edited by JOHN J. REESE, M. D., Professor of Medical Jurisprudence and Toxicology in the University of Pennsylvania. In one large octavo volume of 937 pages, with 70 illustrations. Cloth, $5.00; leather, $6.00; half Russia, raised bands, $6.50.

The American editions of this standard manual have for a long time laid claim to the attention of the profession in this country; and the eighth comes before us as embodying the latest thoughts and emendations of Dr. Taylor upon the subject to which he devoted his life with an assiduity and success which made him *facile princeps* among English writers on medical jurisprudence. Both the author and the book have made a mark too deep to be affected by criticism, whether it be censure or praise. In this case, however, we should only have to seek for laudatory terms.—*American Journal of the Medical Sciences,* Jan. 1881.

This celebrated work has been the standard authority in its department for thirty-seven years, both in England and America, in both the professions which it concerns, and it is improbable that it will be superseded in many years. The work is simply indispensable to every physician, and nearly so to every liberally-educated lawyer, and we heartily commend the present edition to both professions.—*Albany Law Journal,* March 26, 1881.

By the Same Author.
The Principles and Practice of Medical Jurisprudence. Third edition. In two handsome octavo volumes, containing 1416 pages, with 188 illustrations. Cloth, $10; leather, $12. *Just ready.*

For years Dr. Taylor was the highest authority in England upon the subject to which he gave especial attention. His experience was vast, his judgment excellent, and his skill beyond cavil. It is therefore well that the work of one who, as Dr. Stevenson says, had an "enormous grasp of all matters connected with the subject," should be brought up to the present day and continued in its authoritative position. To accomplish this result Dr. Stevenson has subjected it to most careful editing, bringing it well up to the times.—*American Journal of the Medical Sciences,* Jan. 1884.

By the Same Author.
Poisons in Relation to Medical Jurisprudence and Medicine. Third American, from the third and revised English edition. In one large octavo volume of 788 pages. Cloth, $5.50; leather, $6.50.

PEPPER, AUGUSTUS J., M. S., M. B., F. R. C. S.,
Examiner in Forensic Medicine at the University of London.

Forensic Medicine. In one pocket-size 12mo. volume. *Preparing.* See *Students' Series of Manuals,* page 4.

LEA, HENRY C.
Superstition and Force: Essays on The Wager of Law, The Wager of Battle, The Ordeal and Torture. Third revised and enlarged edition. In one handsome royal 12mo. volume of 552 pages. Cloth, $2.50.

This valuable work is in reality a history of civilization as interpreted by the progress of jurisprudence... In "Superstition and Force" we have a philosophic survey of the long period intervening between primitive barbarity and civilized enlightenment. There is not a chapter in the work that should not be most carefully studied; and however well versed the reader may be in the science of jurisprudence, he will find much in Mr. Lea's volume of which he was previously ignorant. The book is a valuable addition to the literature of social science.—*Westminster Review,* Jan. 1880.

By the Same Author.
Studies in Church History. The Rise of the Temporal Power—Benefit of Clergy—Excommunication. New edition. In one very handsome royal octavo volume of 605 pages. Cloth, $2.50. *Just ready.*

The author is pre-eminently a scholar. He takes up every topic allied with the leading theme, and traces it out to the minutest detail with a wealth of knowledge and impartiality of treatment that compel admiration. The amount of information compressed into the book is extraordinary. In no other single volume is the development of the primitive church traced with so much clearness, with so definite a perception of complex or conflicting sources. The fifty pages on the growth of the papacy, for instance, are admirable for conciseness and freedom from prejudice.—*Boston Traveller,* May 3, 1883.

Allen's Anatomy	6
American Journal of the Medical Sciences	3
American System of Gynæcology	27
American System of Practical Medicine	15
An American System of Dentistry	24
*Ashhurst's Surgery	20
Ashwell on Diseases of Women	28
Attfield's Chemistry	9
Ball on the Rectum and Anus	4, 21
Barker's Obstetrical and Clinical Essays,	29
Barlow's Practice of Medicine	18
Barnes' Midwifery	29
*Barnes on Diseases of Women	27
Barnes' System of Obstetric Medicine	29
Bartholow on Electricity	17
Basham on Renal Diseases	24
Bell's Comparative Physiology and Anatomy	4, 7
Bellamy's Operative Surgery	4, 20
Bellamy's Surgical Anatomy	6
Blandford on Insanity	19
Bloxam's Chemistry	9
*Bristowe's Practice of Medicine	14
Broadbent on the Pulse	4, 16
Browne on the Ophthalmoscope	23
Browne on the Throat	18
Bruce's Materia Medica and Therapeutics	11
Brunton's Materia Medica and Therapeutics	11
Bryant on the Breast	4, 21
*Bryant's Practice of Surgery	21
*Bumstead on Venereal Diseases	25
*Burnett on the Ear	24
Butlin on the Tongue	4, 21
Carpenter on the Use and Abuse of Alcohol	8
*Carpenter's Human Physiology	8
Carter on the Eye	23
Century of American Medicine	14
Chambers on Diet and Regimen	18
Charles' Physiological and Pathological Chem.	10
Churchill on Puerperal Fever	28
Clarke and Lockwood's Dissectors' Manual	4, 6
Classen's Quantitative Analysis	10
Cleland's Dissector	5
Clouston on Insanity	19
Clowes' Practical Chemistry	10
Coats' Pathology	13
Cohen on the Throat	18
Coleman's Dental Surgery	24
Condie on Diseases of Children	30
Cornil on Syphilis	25
*Cornil and Ranvier's Pathological Histology	13
Cullerier's Atlas of Venereal Diseases	25
Curnow's Medical Anatomy	4, 6
Dalton on the Circulation	7
*Dalton's Human Physiology	8
Davis' Clinical Lectures	16
Draper's Medical Physics	7
Druitt's Modern Surgery	20
Duncan on Diseases of Women	28
*Dunglison's Medical Dictionary	4
Edes' Materia Medica and Therapeutics	12
Edis on Diseases of Women	27
Ellis' Demonstrations of Anatomy	7
Emmet's Gynæcology	28
*Erichsen's System of Surgery	21
Esmarch's Early Aid in Injuries and Accid'ts	22
Farquharson's Therapeutics and Mat. Med.	12
Fenwick's Medical Diagnosis	16
Finlayson's Clinical Diagnosis	16
Flint on Auscultation and Percussion	18
Flint on Phthisis	18
Flint on Physical Exploration of the Lungs	18
Flint on Respiratory Organs	18
Flint on the Heart	18
*Flint's Clinical Medicine	16
Flint's Essays	16
*Flint's Practice of Medicine	14
Folsom's Laws of U. S. on Custody of Insane	19
Foster's Physiology	8
*Fothergill's Handbook of Treatment	16
Fownes' Elementary Chemistry	9
Fox on Diseases of the Skin	26
Frankland and Japp's Inorganic Chemistry	9
Fuller on the Lungs and Air Passages	18
Galloway's Analysis	8
Gibney's Orthopædic Surgery	22
Gould's Surgical Diagnosis	4, 20
*Gray's Anatomy	5
Greene's Medical Chemistry	10
Green's Pathology and Morbid Anatomy	13
Griffith's Universal Formulary	11
Gross on Foreign Bodies in Air-Passages	18
Gross on Impotence and Sterility	25
Gross on Urinary Organs	24
*Gross' System of Surgery	20
Habershon on the Abdomen	17
*Hamilton on Fractures and Dislocations	22
Hamilton on Nervous Diseases	19
Hartshorne's Anatomy and Physiology	6
Hartshorne's Conspectus of the Med. Sciences	3
Hartshorne's Essentials of Medicine	14
Hermann's Experimental Pharmacology	11
Hill on Syphilis	25
Hillier's Handbook of Skin Diseases	26
Hoblyn's Medical Dictionary	4
Hodge on Women	28
Hodge's Obstetrics	28
Hoffmann and Power's Chemical Analysis	10
Holden's Landmarks	5
Holland's Medical Notes and Reflections	17
*Holmes' System of Surgery	22
Horner's Anatomy and Histology	6
Hudson on Fever	14
Hutchinson on Syphilis	4, 25
Hyde on the Diseases of the Skin	26
Jones (C. Handfield) on Nervous Disorders	18
Juler's Ophthalmic Science and Practice	23
Keating on Infants	30
King's Manual of Obstetrics	29
Klein's Histology	4, 13
Landis on Labor	29
La Roche on Pneumonia, Malaria, etc.	18
La Roche on Yellow Fever	14
Laurence and Moon's Ophthalmic Surgery	23
Lawson on the Eye, Orbit and Eyelid	23
Lea's Studies in Church History	31
Lea's Superstition and Force	31
Lee on Syphilis	25
Lehmann's Chemical Physiology	8
*Leishman's Midwifery	30
Lucas on Diseases of the Urethra	4, 24
Ludlow's Manual of Examinations	3
Lyons on Fever	14
Maisch's Organic Materia Medica	11
Marsh on the Joints	4, 22
May on Diseases of Women	28
Medical News	1
Medical News Visiting List	2
Medical News Physicians' Ledger	3
Meigs on Childbed Fever	28
Miller's Practice of Surgery	20
Miller's Principles of Surgery	20
Mitchell's Nervous Diseases of Women	19
Morris on Diseases of the Kidney	4, 24
Morris on Skin Diseases	26
Neill and Smith's Compendium of Med. Sci.	3
Nettleship on Diseases of the Eye	23
Norris and Oliver on the Eye	23
Owen on Diseases of Children	4, 30
*Parrish's Practical Pharmacy	11
Parry on Extra-Uterine Pregnancy	29
Parvin's Midwifery	29
Pavy on Digestion and its Disorders	18
Pepper's System of Medicine	15
Pepper's Forensic Medicine	4, 31
Pepper's Surgical Pathology	4, 13
Pick on Fractures and Dislocations	4, 22
Pirrie's System of Surgery	20
Playfair on Nerve Prostration and Hysteria	19
*Playfair's Midwifery	29
Politzer on the Ear and its Diseases	24
Power's Human Physiology	4, 8
Purdy on Bright's Disease and Allied Affections	24
Ralfe's Clinical Chemistry	4, 10
Ramsbotham on Parturition	28
Remsen's Theoretical Chemistry	9
*Reynolds' System of Medicine	15
Richardson's Preventive Medicine	17
Roberts on Urinary Diseases	24
Roberts' Compend of Anatomy	7
Roberts' Principles and Practice of Surgery	20
Robertson's Physiological Physics	4, 7
Ross on Nervous Diseases	19
Sargent's Minor and Military Surgery	20
Savage on Insanity, including Hysteria	4, 19
Schäfer's Essentials of Histology,	13
Schreiber on Massage	17
Seiler on the Throat, Nose and Naso-Pharynx	18
Series of Clinical Manuals	4
Simon's Manual of Chemistry	9
Skey's Operative Surgery	20
Slade on Diphtheria	18
Smith (Edward) on Consumption	18
*Smith (J. Lewis) on Children	30
Stillé on Cholera	18
*Stillé & Maisch's National Dispensatory	12
*Stillé's Therapeutics and Materia Medica	12
Stimson on Fractures	22
Stimson's Operative Surgery	22
Stokes on Fever	14
Students' Series of Manuals	4
Sturges' Clinical Medicine	16
Tanner on Signs and Diseases of Pregnancy	29
Tanner's Manual of Clinical Medicine	16
Taylor on Poisons	31
*Taylor's Medical Jurisprudence	31
Taylor's Prin. and Prac. of Med. Jurisprudence	31
*Thomas on Diseases of Women	27
Thompson on Stricture	24
Thompson on Urinary Organs	24
Tidy's Legal Medicine	31
Todd on Acute Diseases	21
Treves' Manual of Surgery	4, 6
Treves' Surgical Applied Anatomy	4, 6
Treves on Intestinal Obstruction	4, 21
Tuke on the Influence of Mind on the Body	19
Visiting List, The Medical News	2
Walshe on the Heart	18
Watson's Practice of Physic	14
*Wells on the Eye	23
West on Diseases of Childhood	30
West on Diseases of Women	27
West on Nervous Disorders in Childhood	30
Williams on Consumption	18
Wilson's Handbook of Cutaneous Medicine	26
Wilson's Human Anatomy	5
Winckel on Pathol. and Treatment of Childbed	28
Wöhler's Organic Chemistry	9
Woodhead's Practical Pathology	13
Year-Books of Treatment for 1884 and 1885	17

Books marked * are also bound in half Russia.

LEA BROTHERS & CO., Philadelphia.

www.ingramcontent.com/pod-product-compliance
Lightning Source LLC
Chambersburg PA
CBHW030740230426
43667CB00007B/791